Dry Mineral Processing

Saeed Chehreh Chelgani · Ali Asimi Neisiani

Dry Mineral Processing

Saeed Chehreh Chelgani
Minerals and Metallurgical Engineering,
Department of Civil, Environmental
and Natural Resources Engineering
Luleå University of Technology
Luleå, Sweden

Ali Asimi Neisiani
Department of Mining and Metallurgical
Engineering
Yazd University
Yazd, Iran

ISBN 978-3-030-93749-2 ISBN 978-3-030-93750-8 (eBook)
https://doi.org/10.1007/978-3-030-93750-8

This Springer imprint is published by the registered company Springer Nature Switzerland AG
The registered company address is: Gewerbestrasse 11, 6330 Cham, Switzerland

To our parents

Preface

This book has been written to fill the gap in the literature by introducing various mineral beneficiation methods in a dry environment. In mineral processing, a wet environment is currently the main condition for the separation. However, based on the United Nations released information, by 2030, over 2 billion people will have unsafe water for drinking, and 700 million people will be displaced or at risk of displacement by intense water shortage. Water scarcity is one of the critical issues that could be the major cause of wars in upcoming years. During the last four decades, rapid developments made the processing of low-grade ores meaningful. By decreasing ore grades, mineral beneficiation plants need to use more water to process more materials and keep production rates. In other words, the ore upgrading plants are highly dependent on water which its scarcity poses a significant risk to all of us in the industry. Thus, water management would be a strategic issue for mineral processing plants since water scarcity has become more common. This has raised the risk of water competition with local communities and farmers and markedly increased the importance of water quality in recycling tailing issues. All these would highlight the importance of dry mineral processing.

This book is dedicated to various mineral separation beneficiation methods which can operate in dry conditions. The main aim is to expand dry processing knowledge and its fundamental aspects and make a foundation for further investigations. The authors would like to thank all those who supported us in gathering information and writing this book.

Luleå, Sweden
2021

Saeed Chehreh Chelgani

Contents

Abbreviations

ABS	Acrylonitrile butadiene styrene
ADMFBS	Air dense medium fluidized bed separator
ASBS	Automatic sensor-based sorting
B	Breakage distribution function
CCD	Color camera detection
CCFC	Counter-current fluidized cascade
d	Particle diameter
DEA	Diethylamine
DE-XRT	Dual-energy x-ray transmission
DHIMS	Dry high-intensity magnetic separators
DLIMS	Dry low-intensity magnetic separators
DMS	Dense medium separation
EM	Electromagnetic
ESP	Electrostatic plate
f	Movement frequency
F_b	Effective buoyancy force
F_f	Frictional force
F_{gd}	Drag force of the air
FGX	Fuhe Ganfa Xuan mei
F_{mg}	Force of gravity
F_{sd}	Drag force of the air dense medium
g	Gravitational constant
G	Gravitational force of the particle
HDMS	Heavy-dense medium separation
HGMS	High-gradient magnetic separator
HPGR	High-pressure grinding roll
HTR	High-tension roll
i	Expansion ratio of the magnetically stabilized fluidized bed
IMRS	Induced magnetic roll separator
IR	Infrared
LIBS	Laser-induced breakdown spectrometry

LIF	Laser-induced fluorescence
LRMS	Lift roll magnetic separator
LT	Laser triangulation
MA	Mechanical activation
MSFB	Magnetically stabilized fluidized bed
MW-IR	Microwave infrared
NdFeB magnets	Neodymium iron boron magnets
NIR	Near infrared spectroscopy
PAA	Polyacrylic acid
PGNAA	Prompt gamma neutron activation analysis
PM	Photometric
P_{max}	Maximum power
PP	Polypropylene
PVC	Polyvinyl chloride
Q_v	Reaction force
R	Lift force
REDS	Rare-earth drum separators
RERS	Rare-earth roll separators
r_m	Mass fraction of fines in the magnetically stabilized fluidized bed
RM	Radiometric
ROM	Run of mine
RTS	Rotary tribo-electrostatic separator
r_v	Volume fraction occupied by the magnetic medium
S	Specific breakage rate function
Sm-Co magnets	Samarium–cobalt magnets
SMD	Stirred media detritors
TEA	Triethylamine
TR	Thermal infrared
v	Velocity of the particle
VIS	Visual spectrometry
v_{mf}	Minimum fluidization velocity
WEEE	Waste electrical and electronic equipment
WPCB	Waste printed circuit boards
XPS	X-ray photoelectron spectroscopy
XRF	X-ray fluorescence
XRL	X-ray luminescence
XRT	X-ray transmission
ρ_b	Density of the magnetically stabilized fluidized bed
ρ_f	Specific density of the medium particle
ρ_h	Specific density of the heavy particle
ρ_l	Specific density of the light particle
ρ_p	Particle density
φ	Angle to horizontal line
ω_{Pmax}	The speed of maximum power

Chapter 1
Grinding

1.1 Introduction

Comminuting raw materials are extensively utilized in various industries such as food, cosmetics, electronics, pharmaceutical, cement, recycling, and mineral processing. Bringing the input material to the desired size is one of the most energy-consuming and costly processes, consuming around 4% of the worldwide electricity production. This process can be performed in wet or dry mode. Comminution is mainly carried out for two reasons: final product preparation, according to the required size, and the liberation of valuable materials/minerals from their matrices. In recycling and mineral processing, the latter reason is defined as a process in which valuable particles were liberated from the gangue texture before being processed [1–4]. In mineral separation processes, wet grinding has always been the most common technique for reducing particle size and liberating them, and different types of mills have been used in the beneficiation plants (Fig. 1.1).

Conventionally, many mineral processing plants conduct ore comminution stage by dry crushing, followed by wet grinding and classification steps to achieve target size. However, water scarcity dictates to limit of the use of water in ore processing plants, particularly in the arid regions [5, 6]. Besides water shortage, other problems may arise in wet grinding, where moisture must be removed from the process product. Moreover, wet grinding can lead to chemical instability, contamination and negatively affect downstream processes. These reasons call for special precautions, such as milling in an inert or dry atmosphere. Consequently, investigation on the dry grinding concept is essential from an economic and environmental perspective [7, 8]. Comparing wet and dry processing shows that the wet method would be preferable (Fig. 1.2) [9]. However, in the case of specific materials, unique characteristics dictate the mode of the grinding process (dry or wet). Therefore, various aspects of wet and dry grinding technologies should be investigated since the grinding process considerably influences economic mineral processing. For instance, wet grinding is the preferable environment for sulfide mineral beneficiations since the downstream processes will be carried out in wet conditions.

S. C. Chelgani and A. Asimi Neisiani, *Dry Mineral Processing*,
https://doi.org/10.1007/978-3-030-93750-8_1

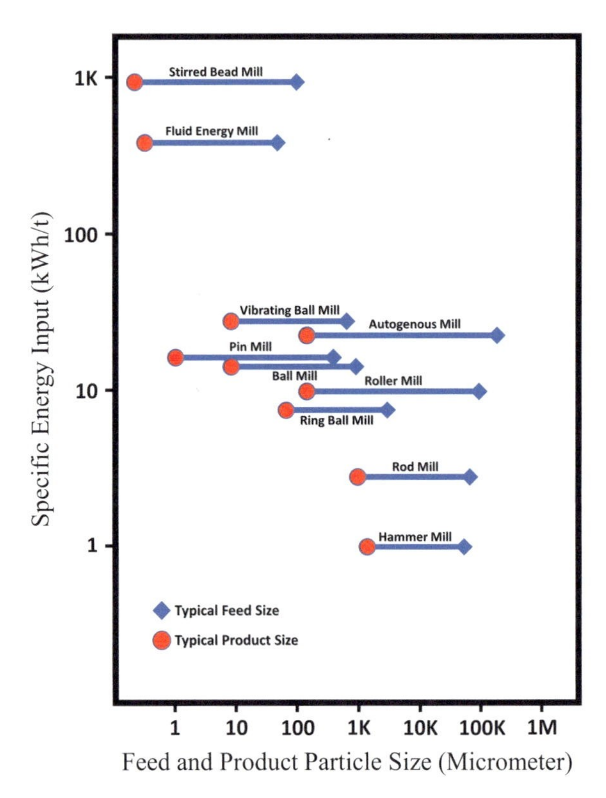

Fig. 1.1 Different mills and their specific energy input values

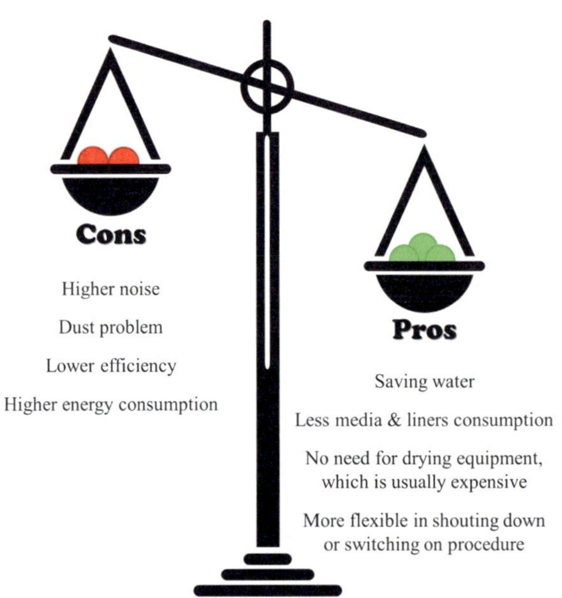

Fig. 1.2 Advantages and disadvantages of dry versus wet grinding in mineral processing

1.2 Process Kinetic

Mechanical comminution is carried out to divide the particles of various materials into smaller pieces by several actions such as breaking, attrition, smashing, crushing, cutting, pealing, and other mechanisms. Figure 1.3 demonstrates three main mechanisms of particle size reduction for mineral beneficiation purposes. The type of size reduction machines is mainly chosen based on the particle size of the feed and the final size requirements [10–12]. Therefore, a combination of different machines (crushers and mills) would be used and considered as a comminution unit in the mineral processing plants.

Grinding kinetics comprises two chief subprocesses, namely, transport and breakage. These two concepts have been repeatedly reported in various references [13–15]. On the other hand, the kinetics of the grinding process could be evaluated using two main models:

- the energy–size relationship model
- the population-balance model.

The population-balance model has surpassed the former due to its ability to describe grinding rates more precisely. Using this model, the entire size distribution with time can be characterized by two main kinetic functions:

- Specific Breakage-rate function (S)
- Breakage-distribution function (B) [16, 17].

Particle breakage could have occurred when the particle is subjected to an efficient breakage action in a grinding zone. In general, the use of the dry or wet technique in grinding makes differences in kinetics. It is a well-known fact that it is challenging and highly time-consuming to achieve ultra-fine particle sizes in the dry environment of a typical tumbling ball mill. This issue appears to happen because of ball coating by fine particles or a bed of fine cohesive materials that cause almost liquid-like properties, making the whole grinding procedure slow and difficult. During these conditions, particles are not exposed to the ball–ball collision action, and inadequate stress is communicated to individual particles for breakage to happen [17, 18].

These factors practically reduce the possibility of particle size reduction process. Therefore, wet mode grinding is more efficient than a dry one and can improve the

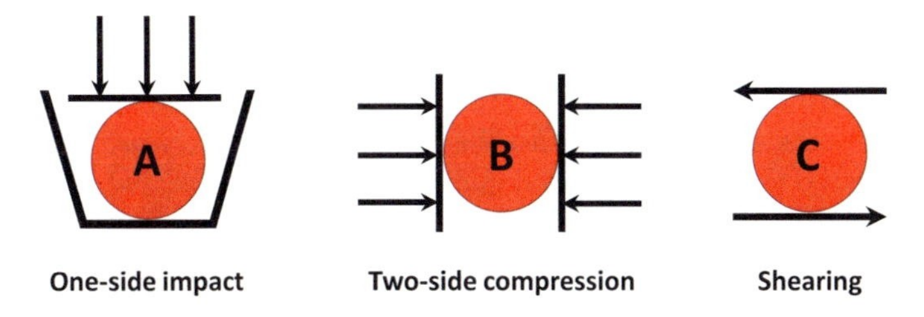

Fig. 1.3 Main mechanisms of particle size reduction [10]

grinding process; however, the improvement level is determined mainly by the type of feed materials [19]. It has been reported that the wet grinding capacity generally is approximately 1.3 times greater than the dry environment under the same industrial operating conditions [19]. Additionally, the specific rate of breakage ratios between dry and wet grinding environments ranges from 1.1 to around 2, depending on the type of material. Water presence in the grinding zone can considerably decrease the slowing down of the process due to washing the coated ball and re-agglomeration of fine particles [17]. The slowing-down factor refers to the ratio of specific breakage rates at a high grinding rate to the normal specific breakage rates. Different materials demonstrate different slowing-down effects since they are varied in cohesive forces. If fine particles accumulate in the milling environment, the slowing-down effect decreases [20].

Moreover, the water improves the transfer of the balls' mechanical force to the particles' stresses, resulting in higher breakage rates. However, a gradual accumulation of fine particles in the milling environment makes the slurry more viscous, and this phenomenon reduces the rate of breakage. The greater the number of fine particles in the grinding zone, the more limiting and sensitive the viscosity influence [21, 22].

In addition, when the slurry concentration is increased markedly, the effective mill diameter is gradually reduced. The coated inner mill wall with a layer of built-up deposits could be the main reason for this phenomenon; thus, the grinding media eventually initiate to stick to the built layer. Consequently, less lifting and tumbling of the grinding media happens (less breakage occurs) [23]. Long-duration dry grinding results in cold-welding and pelletizing of fine particles, whereas the slowing-down phenomenon usually happens relatively quickly. Therefore the pelletization of fine particles could not be demonstrated [17]. Various materials, during dry grinding, show different levels of slowing-down effect, probably because cohesive forces vary from material to material [24].

During grinding, three significant stages of kinetics have been identified (Fig. 1.4):

- *The Rittinger stage*: The particle size is reduced, and grinding time is generally dedicated to a considerable increase of specific surface area. During this step, comminution occurs comparatively rapidly because of many-particle defects such as pores, dislocations, lattice defects, and inclusions. Additionally, the grinding energy is commensurate with the produced specific surface area in this stage.
- *The aggregation stage* is the second step in which a moderation in the size reduction intensity occurs. In this section, interactions between particle-grinding media, particle-liners, and particles with each other are increased. Thus, particle adhesion to each other, grinding media, and the liners begin to occur. Moreover, during this stage, the curve flattens because the energy efficiency decreases. A portion of energy in this section is used for deforming the particle layer, which is adhered to the mentioned surfaces.
- *The agglomeration stage* is the third section, where the dispersion of particles gradually decreases after a long period of grinding. Through these conditions, particles start to be agglomerated after a certain maximum fineness. At

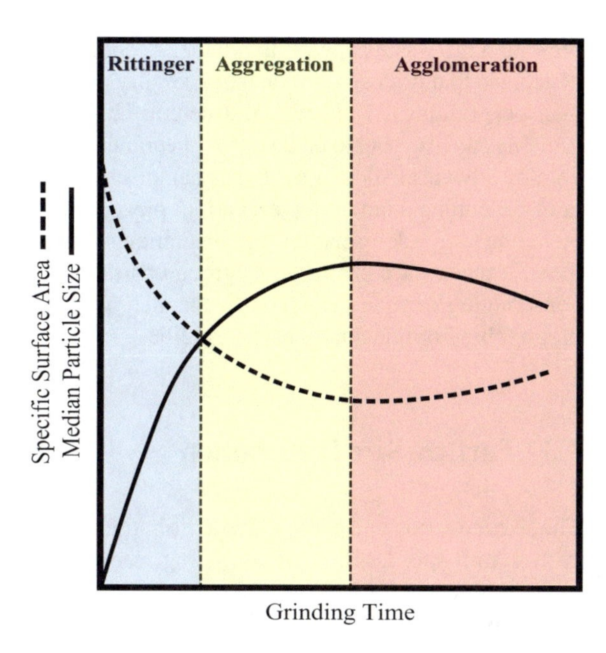

Fig. 1.4 Significant stages of grinding kinetics [27]

this stage, phenomena such as re-organization of crystal structure and mechanochemical changes may occur, showing that appropriate energy is introduced. Also, particle structural alteration or mechanical activation mainly happens in this stage [5, 25–27].

By particle size reduction, the structural defects in the particles and milling resistance have increased. This probably caused that the transition happens from the aggregation stage to the mechanical activation section. Subsequently, the mechanical energy mode in particles changes from breakage to plastic flow, making an intense distortion and dislocations of the crystal structure, producing more defects in particles. These defects in the particles' structure significantly influence the chemical characteristics of the material, leading to mechanochemical activation [28, 29].

Compared to the wet grinding method, dry grinding requires more energy and time to achieve a particular particle size distribution. Consequently, parts of this required greater energy might be used in the form of particle defects. The structural defects could comprise extra enthalpy content even more than the surface energy. For this reason, the extra enthalpy in the samples which are ground in a dry environment is greater than those that are ground in a wet grinding process. These defects can lead to making comparatively rough particle surfaces. Topographical investigations by atomic force and scanning electron microscopy have demonstrated more surface roughness on particles being dry ground than particles ground in wet mode. Therefore, the material surfaces can be extremely activated or deactivated the particle surfaces, leading to a change in their behavior in the post-grinding procedures [30].

An investigation on the dry mechanical activation of mine waste has shown that after 120 min of dry grinding by a planetary mill, a portion of the serpentine present in the waste converts to olivine. According to this phenomenon, the authors concluded that, for carbonation purposes, dry mechanical activation is more suitable than wet grinding [31]. However, structural defects in mineral particles caused by dry grinding can also lead to undesirable phenomena, such as increasing preg-robbing of gold by silicates. Another study has demonstrated that after 30 min dry grinding of gold-containing quartz samples, various physicochemical alteration types can occur, including lattice deformation, agglomeration, amorphization, and surface area increment. If the surface chemistry of ground particles changes, low valence silicon and non-bridging oxygen centers are generated, playing a significant role in the gold preg-robbing of fine quartz particles [29].

1.3 Particle Size Distribution

The size reduction and particle size distribution are influenced by various parameters such as mill speed, filling rate, grinding media size and type, pulp density (in wet mode grinding), dissemination features, material hardness, and grinding environment (dry or wet) [4, 32–36]. Moreover, the feed materials' characteristics and size distribution are very influential, and the grinding environment is not an exception [37, 38]. Producing a broader size distribution of ground particles is one of the main drawbacks of dry grinding compared to the wet technique [20, 37, 39, 40]. It is documented that the median particle size of the product in grinding using a dry ball mill had been approximately four times coarser compared to products produced in wet grinding [37]. The particle size distribution of products from the grinding process is an essential parameter since many downstream processes such as magnetic separation, gravity separation, froth flotation, leaching, and roasting are sensitive to the particle size [4, 41]. A narrow particle size distribution is usually advantageous to improve the performance of downstream processes in mineral processing. It also significantly reduces classification and separation operation costs. Therefore, controlling particle size distribution is an essential factor for dry grinding [37, 41]. Additionally, many applications of various minerals wholly depend on their surface properties and particle size [42–44]. For example, in the case of clay, small-sized particles exhibit their colloidal physicochemical features. This type of clay is appropriate for several ceramic, catalytic, and engineering applications due to its applicability to surface reactivity and large specific surface area [45].

The grinding environment causes a noticeable difference in process kinetics as well as the features of the ground products. As long as the solid concentration in the mill is less than 50%, particle breakage rates are faster for wet grinding than those for the dry environment. This phenomenon is attributable to the interactions between water molecules and broken surface bonds in the wet process, which cannot occur in dry [46, 47]. Moreover, when slurry viscosity is rapidly increased during wet grinding, a layer of particles initiates to accumulate around the inner part of the mill wall. Thus, contact between the particles and grinding media reduces the efficiency

Table 1.1 Summary of dry ultrafine grinding tests carried out with quartz and silica

Grinding Setting	Feed size (μm)	Grinding Period	Grinding limit (μm)	Agglomeration	References
Ball Mill	50	1–90 h	10 μm (25 h)	Yes	[56]
Ball Mill	<90	72–360 h	1.3 μm (360 h)	No	[57]
Oscillating Mill	13	15–600 s	5 μm (10 min)	Yes	[58]
Vibrating Rod Mill	<74	1–48 h	1 μm (24 h)	Yes	[59]
Planetary Ball Mill	<74	10–60 h	0.08 μm (30 h)	Yes	[60]
Jet Mill	16.3 ∓ 1.3	30 min	3.2 μm	No	[61]

of contact between the particles and grinding media, leading to a notable decrease in the grinding level [22, 48].

In the case of ultra-fine grinding, along with the high energy consumption, the greatest problematic issue would be related to particle-particle interactions. This is closely dependent on the grinding system and the surface reactivity. This problem is critical, specifically in dry grinding processes, where the minimum particle size of the ground materials is influenced by two contradictory effects of particle agglomeration and particle breakage mechanisms [49, 50]. When the agglomeration mechanism is presented, the minimum attained particle size would be recognized as the grinding limit [51, 52]. When this level is achieved, it has been advocated that a considerable portion of energy be delivered by the grinding media. This energy is absorbed by the plastic deformation of crystalline particles, which characterize a typical brittle-ductile transition [53–55]. This issue can be observed in Table 1.1, which demonstrates a brief summary of some investigations related to dry ultrafine grinding of quartz and silica particles.

1.4 Energy Consumption

The grinding operation is the most costly stage in mineral processing. Figure 1.5 demonstrates relative costs for a typical copper concentrator plant. It can be seen that approximately half of the mineral processing costs are related to the grinding process and its energy consumption [62]. Consequently, understanding the relationship between the grinding operating conditions and power requirements is worthy of attention throughout the new grinding machine's design and during its operation.

Regarding a particular particle size distribution, investigations indicate that in dry type grinding, the energy requirement is approximately 15–50% higher than in a wet mode [3, 30, 39, 63]. Using an electromagnetic mill, a comparison between these two grinding environments has shown; while wet grinding requires around 1.6 kW

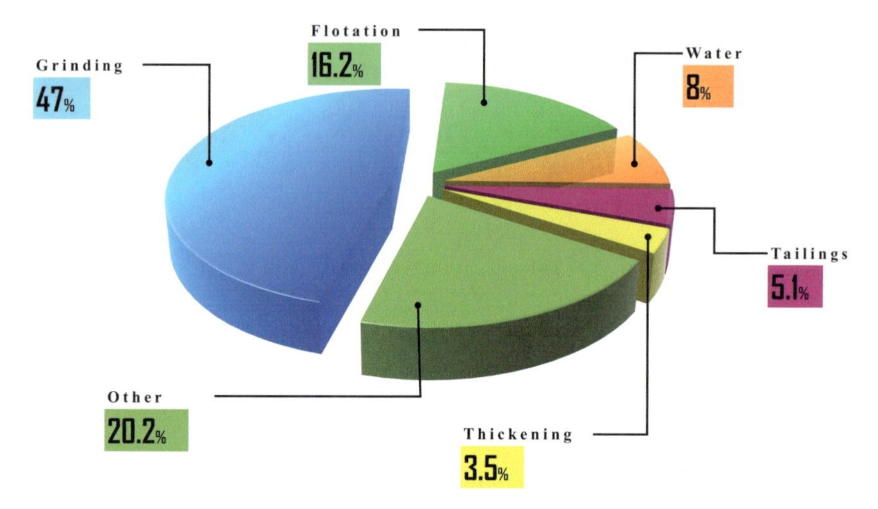

Fig. 1.5 Relative costs for a typical copper concentrator [62]

of energy, the dry condition needs 3 kW for the same condition. It also has been shown that for an equal level of energy, the size reduction ratio in the wet method is higher than dry grinding [10]. As mentioned, during dry grinding, a portion of this excess energy leads to a structural defect in mineral particles. This might result in the mechanical activation of particles. Nevertheless, in wet grinding, slight structural deformation and new surfaces can be expected [30].

Castro & Valenzuela have conducted two investigations to show the effect of water content on the amount of energy required in grinding and the difference between the amount of energy needed for the dry and wet environments [64, 65]. Power-speed curves presented in Fig. 1.6 depict the effect of different amounts of water in the mill. These curves show that the water content has two significant effects on the power requirement throughout wet mode grinding. The first effect is that when the water

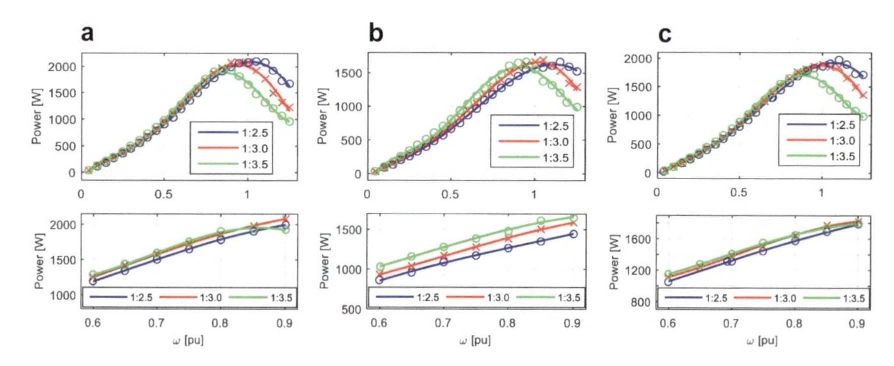

Fig. 1.6 Power-speed curves for various water-to-gravel ratios. Top row: Total speed interval; Bottom row: Speed interval 0.6–0.9 critical speed [65]

content in the mill increases, the speed of maximum power (ω_{Pmax}) also increases. Additionally, along with the increase in the maximum power speed, there is also an increase in the maximum power (P_{max}). The increase in (ω_{Pmax}), when a higher amount of water is present in the grinding zone, is described by a lower density of the load kidney, such as the liners and lifters. This phenomenon leads to dragging thinner charge layers nearer to the mill interior surface. The consequence is that the mill charges initiate to centrifuge at a higher speed. Due to water content in the wet grinding process, the critical speed for tests A and B (lower ball percentages) happens in the speed range 0.6–0.8. Based on these curves, in the mentioned speed range, the power requirement is lower for a greater amount of water when the charge condition (balls and gravel percentages) is constant. This phenomenon is especially observed for tests A, in which this situation remains visible up to 0.9 of critical speed [64, 65].

Additionally, it is anticipated that the lower the density of the load kidney, the closer to the internal mill surface the grinding media would be. Thus, a higher contribution to the torque demand is expected. Nonetheless, two other affecting factors lead to power consumption reduction in a lower density process. First, the presence of a larger amount of water decreases the viscosity of slurry; therefore, a lower percentage of the grinding load is elevated by lifters. And secondly, when the water content in the grinding zone is greater than the free space volume between gravels and balls (because of their porosity), the so-called pool effect occurs. This effect decreases the power requirement for a specific operating condition (charge and speed) [66–68].

Comparing dry and wet grinding curves in a ball mill in the same condition (an equal amount of materials and grinding media has been used on each pair of wet-dry experiments) (Fig. 1.7) indicated two significant differences at higher power values of wet grinding, specifically when speed exceeded 0.6 critical speed and for higher ω_{Pmax} in the wet grinding. The only distinction was that water had been added to the system (along with materials and grinding media). Although the mass of the added water should cause an increase in the power requirement for the wet grinding experiments, the observed difference cannot be attributed to the effect of water mass alone. The increase in power reveals that the same charge of grinding media impacts the power demand during wet grinding (compared to dry grinding). Only a minor proportion of this effect is attributable to water mass. The increase in the power demonstrates that the grinding media distribution inside the load kidney is nearer to the interior surface of the mill during wet grinding. Thus, the larger radius to the axis of rotation causes an increase in the load torque.

On the other hand, the increase in ω_{Pmax} suggests that the mill charge initiates to centrifuge at higher speeds for wet grinding. This phenomenon is because of the water present in the system, leading to a lower amount of dragging force between neighbor charge layers than dry grinding. In addition, the comparison of ω_{Pmax} during dry and wet grinding demonstrates that for effective comminution, in the wet grinding process, the operating speed needs to be greater than in dry mode [65].

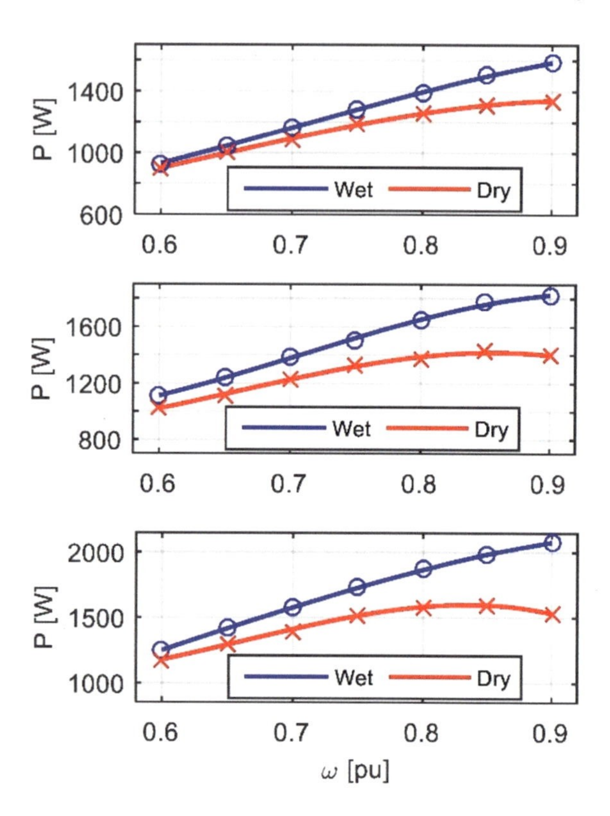

Fig. 1.7 Power curves for dry and wet grinding (water-to-gravel ratio 1:3) [65]

1.5 Downstream Processes

Flotation beneficiation is the extensively used technique for fine particles (+25 to 100 μm) separation [69, 70]. The grinding process, which provides the flotation input particles, has an essential impact on flotation efficiency since it affects factors such as pulp chemistry, particle size distribution, particle surface properties, and mineral liberation [39, 71–73]. The higher energy consumption in the dry system, resulting in mineral structural defects, may change flotation kinetics compared to wet grinding. It means, for dry ground minerals, the kinetics of flotation might be higher, and lower energy consumption is required (based on the mineralogy) [30].

Several studies have stated that the degree of dispersion and variations of the surface energy of particles is the primary factor in comparing wet and dry grinding methods [74–76]. It is indicated that a portion of the energy converts to the surface energy during a wet grinding process when the solid percentage increases. The surface energy is higher in materials that have been ground in a dry environment compared to wet ground ones [30]. Comparison of the probability of particle fracture during wet and dry grinding systems has been shown that when the energy input is constant, the length of surface cracks is larger during wet grinding [37]. It also has been indicated that the values of primary breakage distribution (B) in the wet and dry grinding are approximately equal, while the specific rates of breakage (S) is higher in a dry grinding [20]. The particular rate of breakage ratio between dry and wet grinding varies from 1.1 to 2.0 for various minerals [77].

In the case of gold concentrates containing hematite, dry grinding can greatly enhance the gold leaching rate in the thiosulfate system. Gold leaching can reach from 35% (wet grinding) to 85% (dry grinding) in 8 h (Fig. 1.8). The lower recovery of gold after wet grinding could be related to the goethite formation during milling, which deteriorates the leaching of gold. This formation did not observe during the dry grinding. It has been indicated by the X-ray diffraction analysis results (Fig. 1.9), a portion of the hematite present in concentration is converted into goethite during wet grinding, as shown in Eq. 1.1. Additionally, X-ray photoelectron spectroscopy (XPS) analysis and element content analysis can demonstrate that the material that is

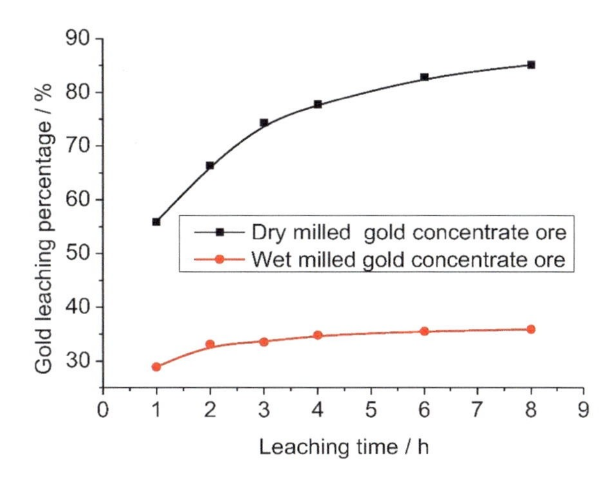

Fig. 1.8 Gold leaching percentage of concentrate calcine. 50 g gold ore in a solution of 10 mM Cu(II), 1.0 M NH_3, and 0.2 M thiosulfate [78]

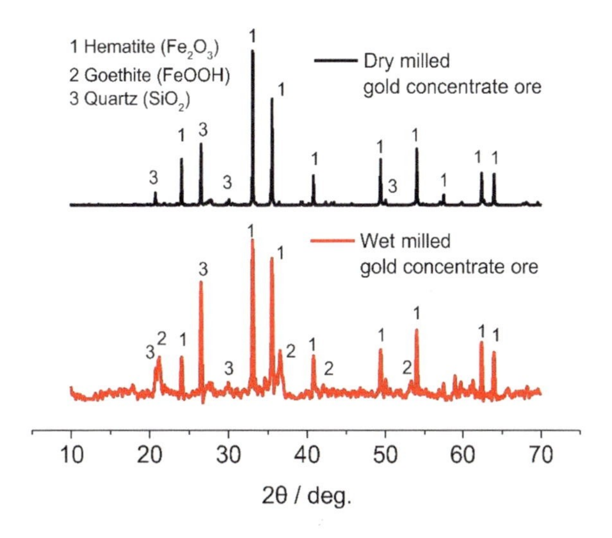

Fig. 1.9 XRD analysis results for leaching residues [78]

adsorbed on the surface of gold particles is mainly goethite, as observed in Figs. 1.10 and 1.11, respectively [78]. This is because this layer covers the gold surface and accelerates thiosulfate decomposition [79, 80]. It was also reported that dry grinding

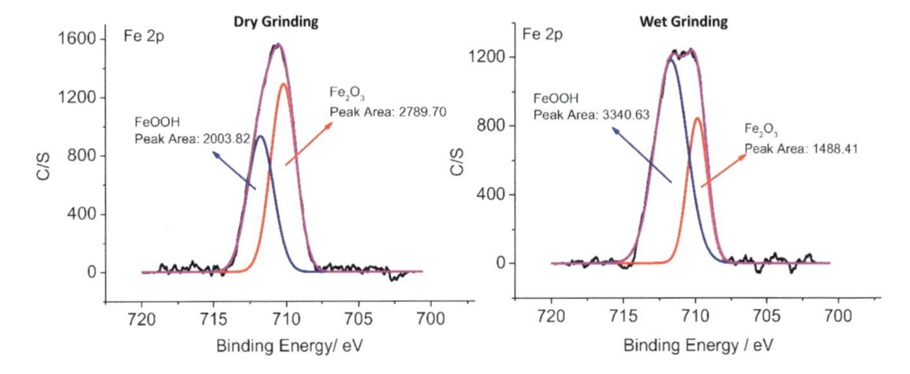

Fig. 1.10 XPS spectra of Fe 2p on the gold surface under different grinding conditions [78]

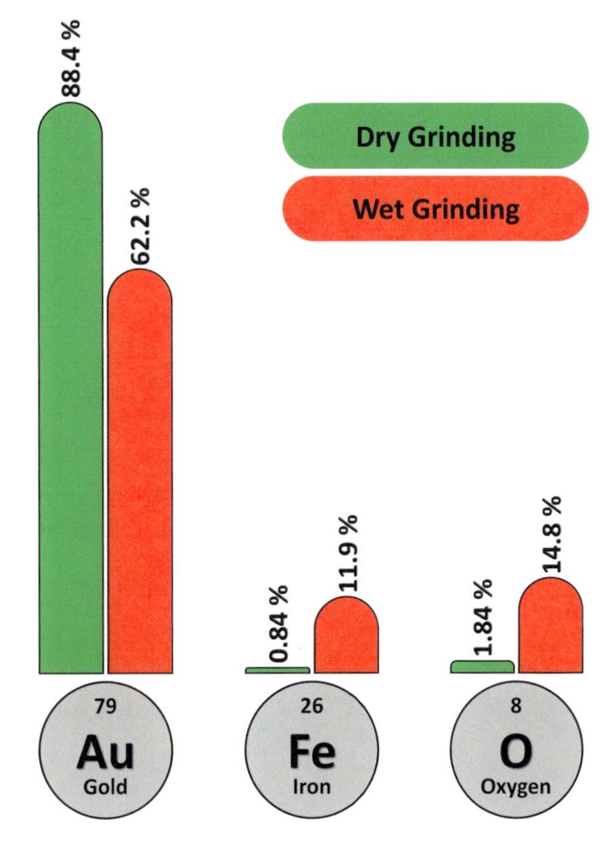

Fig. 1.11 Element content on the gold surface under different grinding conditions [78]

techniques could mechanically activate K-feldspar and improve potassium leachability. High intensive dry grinding of K-feldspar can significantly increase surface area, deform its lattice structure, and improve potassium recovery [81].

$$Fe_2O_3 + H_2O \rightarrow 2FeOOH \tag{1.1}$$

1.6 High-Pressure Grinding Roll (HPGR)

HPGR, as an energy-efficient dry comminution machine, was first utilized industrially in 1985. This technology has considerable advantages such as reducing water consumption, energy requirement, steel usage cost, and increasing capacity. That is why HPGR has gained a special place among size reduction machines used in mineral processing. For example, in the cement production procedure, dry tumbling mills such as rod and ball mills have been used to grind limestone, slag, and clinker over the years. Recently, however, the HPGR has replaced many of the traditional dry grinding systems in this industry [5, 82, 83]. HPGR is equipped with an inter-particle breakage mechanism system in which materials are broken, crushed, and ground by compression. This technology is able to process a broad range of coarse crushing to very finely ground particles [5]. The breakage mechanism in HPGR leads to microcracks along mineral boundaries of materials, resulting in preferential liberation [84, 85]. HPGR can be utilized in both closed and open circuits by using screens or air classification. It was documented that using an HPGR arranged in the closed-circuit can improve the grinding process. The main advantages of this type of HPGR are enhancing throughput by 25%, increasing comminution efficiency, and reducing the specific energy consumption [86]. In the cement industry, this system provides an appropriate feed for next mills or even final high-quality product around 25 μm [6].

One of the most important advantages of HPGR is its ability to save energy. Using HPGR instead of conventional tumbling mills can reduce energy consumption by around 10–50% [84, 87]. An investigation conducted on magnetite ore grinding with a dry HPGR machine to reduce the particle size from 50 mm to 90 μm has revealed that using two stages HPGR closed circuit with an air classifier can reduce energy consumption by 46% compared to a circuit in which there is a tertiary crushing with a wet ball mill. Additionally, the wet grinding process produces slurry, which needs thickening and filtration steps, increasing the plant capital cost [6].

1.7 Grinding Aids

As mentioned, the dry grinding method has recently become an industrially extensively utilized process due to water shortages. However, this technique also has drawbacks compared to wet grinding, such as higher specific energy consumption, the occurrence of agglomeration phenomenon, broader particle size distribution, which can have an adverse effect on post-grinding processes [5, 88]. Generally, grinding process efficiency can mainly be improved by either improving the grinding behavior of the material being ground or enhancing the grinding machines. Recently, the use of additives so-called grinding aids, which is applied in a small amount of 0.01–0.25 wt% related to the mass of product, has reduced these problems and been satisfactorily effective [89]. The grinding aids could be either inorganic or organic; however, the latter is more common commercially [90]. They mainly decrease the interaction between ground particles during grinding and enhance the flowability and dispersibility of the product powder (Fig. 1.12) [91].

Of course, it should be mentioned that these additives have to be selected with particular regard to a special grinding system to be effective, meaning the resulting flow characteristics of the product powder and also the existing grinding circumstances inside the mill. The influence of grinding aids greatly depends on the chemical nature of these additives [49]. The main effects of grinding aids on the dry grinding system can be observed in Fig. 1.13.

During industrially fine and ultrafine grinding, grinding aids are added to the milling process in order to improve product throughput, reduce specific energy consumption, and reach a particular product fineness [49]. Sometimes, these additives are utilized to facilitate material handling and improve product features such as settling times and cement strength [92, 93]. Several investigations have been conducted to investigate the effect of grinding aids on the comminution of various minerals such as cement [94], feldspar [89], limestone [95], quartz [59], etc. It is reported that grinding aids improve the grinding process by altering the adhesive forces, the state of fine particle dispersion, and powder flow characteristics [49]. Additionally, some research has proven that adding grinding aids can decrease fine particles' tendency to become agglomerated [96]. Since the agglomeration phenomenon can have an adverse effect on milling performance and efficiency, the use of grinding aids can cause particles to be ground to smaller sizes by consuming a certain amount of energy [97]. Furthermore, better particle dispersion occurs due to the decreased adhesion effect and thus the reducing agglomeration phenomenon. Therefore, the classification step improves because fewer agglomerated particles are mistakenly classified and led back to the chamber [98].

Some studies have pointed to the effect of grinding aids on various factors such as energy consumption, agglomeration, product coating on grinding media, specific surface energy, product fineness, particle size distribution, surface morphology, and zeta potential. However, grinding aids' effect on energy consumption reduction has received more attention [49]. As the amount of grinding aid used increases, energy consumption reduction increases to a certain value (maximum level), but further addition does not affect [91, 99]. Table 1.2 shows the effect of grinding aids on energy consumption [90].

Not only the type of grinding aids but also their dosage and other parameters affect the milling process. For example, it is proven that milling performance is a function of additive type, polymer molecular weight, and dosage [95]. The use of polyacrylic acid (PAA) improved the grinding performance. The optimum amount of PAA for the

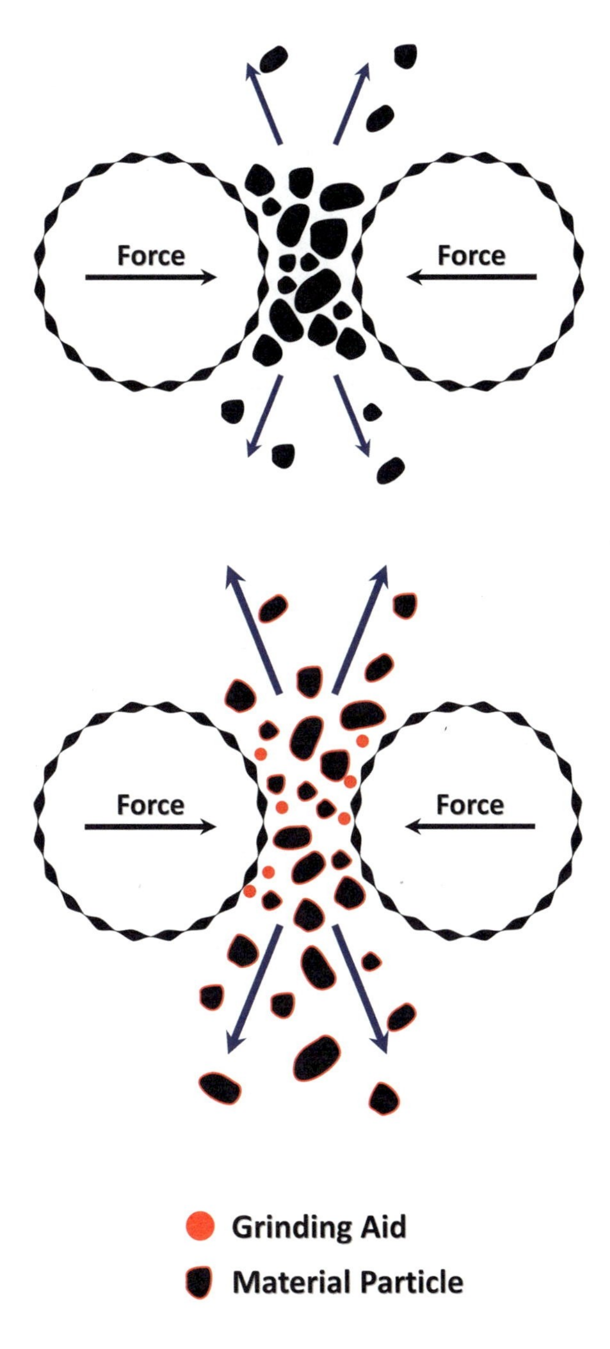

Fig. 1.12 The effect of grinding aids on dispersibility [91]

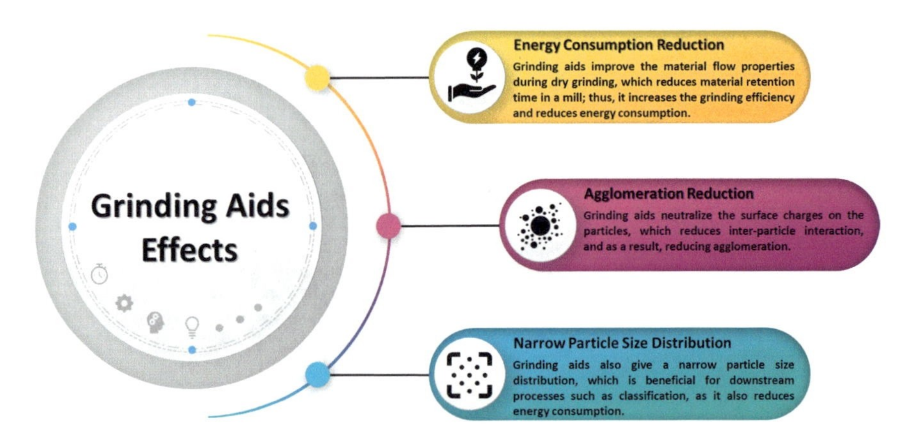

Fig. 1.13 Main effects of grinding aids on the dry grinding system [90]

Table 1.2 Effect of grinding aids on energy consumption [90]

Grinding aid	Material	Reduction in power consumed (%)
Triethylamine (TEA)	Limestone	98.5
BMA-1923™ (Amine based)	Feldspar	60.0
TIPA	Cement	20.6
Propylene glycol	Clinker	10.0
Sodium polyphosphate	Copper ore	15.7
Alcohol	Coal	2.37
Diethylamine (DEA)	Coal	1.05
Sodium polyphosphate	Talc	33.7
Sodium polycarboxylic acid	Talc	20.7
Polyacrylic acid	Calcite	37.2
Triethylamine (TEA)	Cement	17.34
Polyacrylic acid	Limestone	100
Aluminum chloride	Coal	25.0
Sodium silicate	Chromite	4.67

mentioned system was reported to be 0.1%, where energy consumption is minimized. Using more than this amount of grinding aid has had a negative effect on energy consumption. The optimum molecular weight of PPA was 5000 for the grinding of limestone. Higher molecular weight led to an increase in energy consumption. The reason is that an excessive increase in the molecular weight of the polymer can cause particle flocculation [95].

1.8 Stirred Mills

The energy consumption of conventional grinding machines, such as tumbling mills, dramatically increases when the product size below 75 μm is required (Fig. 1.14) [100]. On the other hand, due to the increasing demand of various industries for ultrafine particles, the coarse and intermediate sizes are unsuitable for most industrial applications [91]. In order to produce ultrafine products and liberate valuable material from the gangue mineral, stronger energy intensity and more efficient methods are required. The stirred mill was introduced by Klein and Szegvary in 1928 [101]. Because of its benefits, such as higher efficiency and larger capacity, it has been extensively employed in mineral processing since the 1950s to produce fine and ultrafine ($d80 < 10$ μm) particles. Additionally, because of high energy density, this technology is suitable for mechanical activation (MA) as well [91, 102]. To meet the industrial requirement, both vertical and horizontal types of this machine are available, and the process variables determine the suitable type of grinding. This grinding method is usually applied in wet mode, in which water plays a lubricant role in the milling environment, leading to improving energy efficiency. Additionally, water controls dust generation and decreases air pollution [103].

Many investigations have been carried out regarding various aspects of this relatively new technology [104–108]. Many types of this machine have been designed and developed by manufacturers and utilized for the industrial process. Figure 1.15 shows vertical stirred mills and the pattern of media flow during grinding [103].

A stirrer that rotates at high speed supplies driving force and moves the particles in the milling zone. The feed enters a cylindrical chamber which is usually lined with polyurethane to diminish the abrasion. Unlike tumbling mills, stirred mill's chamber is usually fixed, and an agitator driven by a motor moves the particles and grinding media. Depending on the type and characteristic of input feed, ceramic or zircon grinding media can be utilized [109, 110].

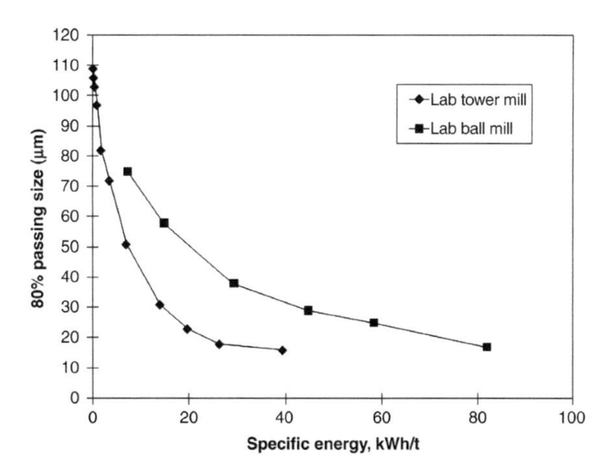

Fig. 1.14 Comparison of grinding efficiency using a Ball Mill and Stirred Mill [100]

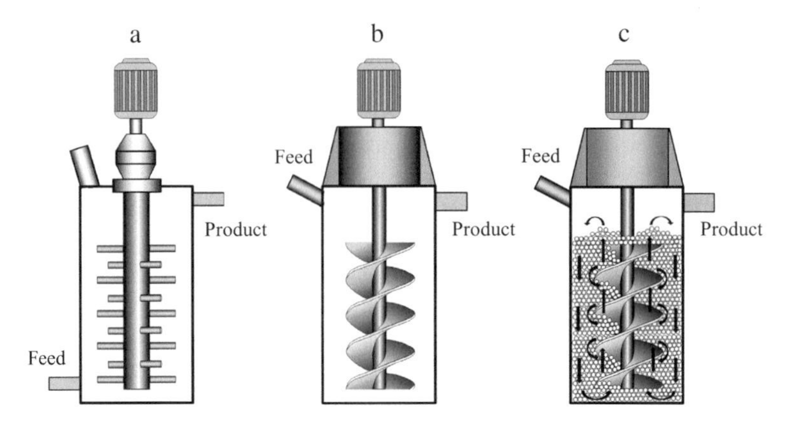

Fig. 1.15 Diagrams of vertical pin (**a**) and vertical screw stirred mill (**b**), and media flow pattern during stirred milling (**c**) [103]

Some investigations demonstrate that stirred mills (compared to ball mills) have better performance when grinding occurs on a micron-scale [110]. This improvement is because the mechanism of supplying energy during stirred milling is different from what happens in ball mills. In ball mills, the abrasive and impact stress lead to particle breakage. These forces are generated when particles and grinding media collide with each other. As the mill rotates, the grinding media are lifted and collide with the mineral particles in a free-fall situation. The maximum force supplied throughout this mechanism almost equals the weight of the grinding medium [110, 111].

Since the revolving part is the mill body, the critical speed limits the mill rotational speed. If mill rotation speed exceeds the critical speed, centrifugal force causes particles and balls to adhere to the mill's inner wall. Therefore, the grinding media do not fall on the mineral particles, and grinding does not occur. Generally, the critical rotational speed in ball mills is around 4 m/s [101]. However, the mechanism applied in stirred milling overcomes this limitation. The stirrer speed in stirred mills can even reach 20 m/s. In this kind of circumstance, the interaction between mineral particles and grinding media is high. Thus, the collision intensity in the stirred mill is much more than in the ball mill. Because of the intensive collision, this method is preferable for producing fine and ultra-fine products, which need more specific energy [101]. It is reported that for achieving a certain fineness, the stirred mill method can reduce energy consumption up to 35% in comparison to conventional types such as tumbling mills [110, 111].

The parameters affecting the Stirred milling process are divided into four general categories.

1. The operating variables of the milling process, such as grinding media, process time, tip speed, etc.
2. The mill operation mode including continuous grinding or batch grinding or circuit.

3. Physical and chemical characteristics of the input feed, namely suspension viscosity.
4. Mill geometry (size and shape) and impeller shape (pin, disc, and screw) [112].

Based on tip speed, which refers to the tangential velocity of the stirrer tip, this type of mills can be categorized into two subgroups:

1. low-speed (around 3 m/s)
2. high-speed (more than 15 m/s) [113].

When the impeller's rotational speed increases, high-energy intensity increases the probability of collision between particles and grinding media. Many investigations have reported that the more the impeller's speed, the finer the products, although energy consumption also increases [114–117].

Eirich Tower Mill and Metso Vertimill are two well-known stirred mills operating at low speed. In this type of machine, the gravitational force is still a key factor. It is possible to decrease the particle size of products to about 15 microns in this method. On the other hand, high-speed stirred mills, in which the rotational speed of the stirrer tip is usually higher than 15 m/s, are able to produce 5 μm or even finer products. Xstrata Isamill and Mesto Stirred Media Detritor (SMD) can be mentioned as high-speed stirred mills. Table 1.3 shows various industrial stirred mills used in mineral processing [62].

Another important parameter affecting the performance and efficiency of stirred mills is grinding media. The grinding media size should be 2–10 mm. They are usually made of zircon-based compounds, ceramic materials, or aluminum oxide in

Table 1.3 Industrial stirred mills are used in mineral processing [62]

Specifications	Mill Name				
	Isa Mill	VertMill TowerMill	SMD (Stirred Media Detritor)	VXP Mill	HIG Mill
Manufacturer	Glencore	Metso Eirich	Mesto	FLSmidth	Outotec
Orientation	Horizontal	Vertical	Vertical	Vertical	Vertical
Impeller shape	Disc	Screw	Pin	Disc	Disc
Speed	High	Low	Medium	Medium	Medium
Impeller tip speed (m/s)	19–23	<3	3–8	10–12	8–12
Density of solids (v/v)	10–30	30–50	10–30	10–30	10–30
Grinding media size (mm)	1–3	12–38	1–8	1.5–3	1–6
Feed size (μm)	300–70	6000–800	300–70	300–70	300–70
Product size (μm)	<10	20	<10	<10	<10
Power intensity (kW/m^3)	300–1000	20–40	50–100	240–765	100–300

order to reduce wear. 50–80% of milling chamber volume is filled by grinding media, and the media are fluidized when impeller rotation reaches an optimum value. The motion of the grinding media inside the chamber wholly depends on stirrer geometry and can be different. The breakage mechanism in a stirred milling process depends on the shearing force originating from the intensive interaction between grinding media and mineral particles [103, 116].

Such as milling methods, in vertical milling, feed rate plays a significant role. Since the grinding specific energy is a function of feed rate and related to the product fineness, a direct correlation between product fineness and feed rate can be observed [103]. When the feed rate decreases, an increase in the product's surface area occurs, and therefore, an inverse relationship between feed rate and product size can be defined [114, 115].

Although stirred milling methods were usually employed in wet comminution applications, the dry mode has recently received special attention. As an important industry that employs dry grinding methods, the cement industry is the 3rd largest energy consumer among various industries. It also emits around 7% of global carbon dioxide emissions, and in this respect, it is in second place. That is why Innovative, efficient technologies that can reduce energy consumption in this industry are very important [118]. Many investigations have been carried out on cement grinding using stirred mills [118–120]. It has been concluded that using stirred mills can reduce energy consumption by 7–18% [118].

MaxxMill, for instance, is a stirred media mill that has successfully been utilized in dry mineral processing [121]. The main parts of a MaxxMill are a revolving chamber, stationary charge deflector, feed and product discharge pipes, and one or more eccentric impeller(s). The impeller's rotation may be in the same or opposite direction of the milling chamber. The chamber can be filled with particles and grinding media (2–10 mm) up to 90% of the total volume of the milling chamber. Optimal grinding media size is determined based on input feed and desired product size. After reaching the desired size, mineral particles continuously come out of the chamber through a pipe installed in the upper part of the mill. In this step, grinding media are not extracted because of the gravitational force. Figure 1.16 shows a simplified diagram of MaxxMill [114, 121].

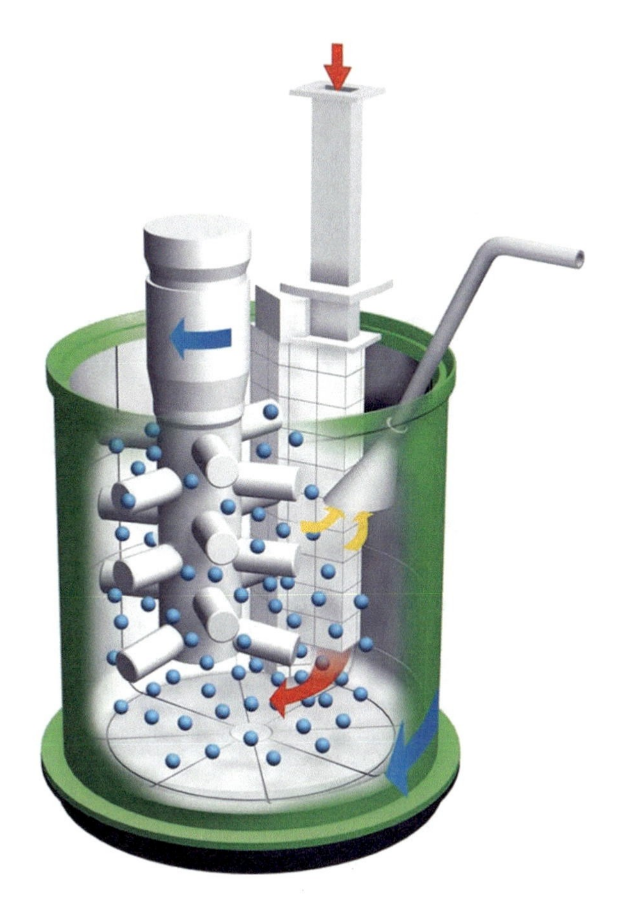

Fig. 1.16 Simplified diagram of MaxxMill [121]

References

1. Fuerstenau, D.W., Abouzeid, A.Z.: The energy efficiency of ball milling in comminution. Int. J. Miner. Process. **67**, 161–185 (2002). https://doi.org/10.1016/S0301-7516(02)00039-X
2. Clermont, B., De Haas, B.: Optimization of mill performance by using online ball and pulp measurements. J. S. Afr. Inst. Min. Metall. **110**, 133–140 (2010)
3. Kotake, N., Kawaguchi, T., Koizumi, H., Kanda, Y.: A fundamental study of dry and wet grinding in bending tests on glass—effect of repeated impact on fracture probability. Miner. Eng. **17**, 1281–1285 (2004). https://doi.org/10.1016/j.mineng.2004.06.030
4. Deniz, V.: Comparisons of dry grinding kinetics of lignite, bituminous coal, and petroleum coke. Energy Sources, Part A Recov. Util. Environ. Eff. **35**, 913–920 (2013). https://doi.org/10.1080/15567036.2010.514591
5. Chelgani, S.C., Parian, M., Parapari, P.S., Ghorbani, Y., Rosenkranz, J.: A comparative study on the effects of dry and wetgrinding on mineral flotation separation—a review. J. Mater. Res. Technol. **8**, 5004–5011 (2019). https://doi.org/10.1016/j.jmrt.2019.07.053
6. Jankovic, A., Suthers, S., Wills, T., Valery, W.: Evaluation of dry grinding using HPGR in closed circuit with an air classifier. Miner. Eng. **71**, 133–138 (2015). https://doi.org/10.1016/j.mineng.2014.10.023

7. Napier-Munn, Timothy J., Morrell, S., Morrison, Robert, D., Kojovic, T.: Mineral comminution circuits: their operation and optimisation (1996)
8. Deniz, V.: A study on the specific rate of breakage of cement materials in a laboratory ball mill. Cem. Concr. Res. **33**, 439–445 (2003). https://doi.org/10.1016/S0008-8846(02)00976-6
9. Bruckard, W.J., Sparrow, G.J., Woodcock, J.T.: A review of the effects of the grinding environment on the flotation of copper sulphides. Int. J. Miner. Process. **100**, 1–13 (2011). https://doi.org/10.1016/j.minpro.2011.04.001
10. Ogonowski, S., Wołosiewicz-Głab, M., Ogonowski, Z., Foszcz, D., Pawełczyk, M.: Comparison of wet and dry grinding in electromagnetic mill. Minerals **8**, 138 (2018). https://doi.org/10.3390/min8040138
11. Ogonowski, S., Ogonowski, Z., Swierzy, M.: Power optimizing control of grinding process in electromagnetic mill. In: Proceedings of the Proceedings of the 2017 21st International Conference on Process Control, PC 2017, pp. 370–375. Institute of Electrical and Electronics Engineers Inc. (2017)
12. Wołosiewicz-Głąb, M., Foszcz, D., Ogonowski, S.: Design of the electromagnetic mill and the air stream ratio model. In: Proceedings of the IFAC-PapersOnLine, vol. 50, pp. 14964–14969. Elsevier B.V. (2017)
13. Abouzeid, A.Z.M.: Transport of particulates in mineral processing systems-tumbling units, pp. 255–262 (1989)
14. Abouzeid, A.Z.M.: Mixing and demixing of particulate solids (1989)
15. Abouzeid, A.Z.M., Fuerstenau, D.W.: Effect of humidity on mixing of particulate solids. Ind. Eng. Chem. Process Des. Dev. **11**, 296–301 (1972). https://doi.org/10.1021/i260042a025
16. Rajamani, R.K., Guo, D.: Acceleration and deceleration of breakage rates in wet ball mills. Int. J. Miner. Process. **34**, 103–118 (1992). https://doi.org/10.1016/0301-7516(92)90018-R
17. Austin, L.G., Klimpel, R.R.P.T.L.: Process Engineering of Size Reduction. SME/AIME, New York (1984)
18. El-Shall, H., Somasundaran, P.: Physico-chemical aspects of grinding: a review of use of additives. Powder Technol. **38**, 275–293 (1984). https://doi.org/10.1016/0032-5910(84)85009-3
19. Bond, F.C.: Crushing and grinding calculations. Part 1. Process. Mach. **1**, 378–385 (1961)
20. Ozkan, A., Yekeler, M., Calkaya, M.: Kinetics of fine wet grinding of zeolite in a steel ball mill in comparison to dry grinding. Int. J. Miner. Process. **90**, 67–73 (2009). https://doi.org/10.1016/j.minpro.2008.10.006
21. Tangsathitkulchai, C.: The effect of slurry rheology on fine grinding in a laboratory ball mill. Int. J. Miner. Process. **69**, 29–47 (2003). https://doi.org/10.1016/S0301-7516(02)00061-3
22. Klimpel, R.: Laboratory studies of the grinding and rheology of coal-water slurries. Powder Technol. **32**, 267–277 (1982). https://doi.org/10.1016/0032-5910(82)85028-6
23. Austin, L.G., Yekeler, M., Dumm, T.F.R.H.: The kinetics and shape factors of ultrafine dry grinding in a laboratory tumbling ball mill (1990)
24. Austin, L.G., Bagga, P.: An analysis of fine dry grinding in ball mills. Powder Technol. **28**, 83–90 (1981). https://doi.org/10.1016/0032-5910(81)87014-3
25. Properties of the grinding product shows that the power consumed for grinding. Change **7**:1–7 (1977)
26. Juhász, A.Z., Opoczky, L.: Mechanical activation of minerals by grinding, pulverizing and morphology of particles. Akadémiai Kiadó 9–234 (1990)
27. Mucsi, G.: A review on mechanical activation and mechanical alloying in stirred media mill. Chem. Eng. Res. Des. **148**, 460–474 (2019). https://doi.org/10.1016/j.cherd.2019.06.029
28. Baláž, P.: Mechanochemistry and nanoscience. In: Mechanochemistry in Nanoscience and Minerals Engineering, pp. 1–102. Springer, Berlin (2008)
29. Mohammadnejad, S., Provis, J.L., Deventer, J.S.J.V.: Effects of grinding on the preg-robbing potential of quartz in an acidic chloride medium. Miner. Eng. **52**, 31–37 (2013). https://doi.org/10.1016/j.mineng.2013.03.003
30. Feng, D., Aldrich, C.: A comparison of the flotation of ore from the Merensky Reef after wet and dry grinding. Int. J. Miner. Process. **60**, 115–129 (2000). https://doi.org/10.1016/S0301-7516(00)00010-7

31. Jiajie Li, M.H.P.: Ultra-fine grinding and mechanical activation of mine waste rock using a planetary mill for mineral carbonation. Int. J. Miner. Process. (2016). https://doi.org/10.1016/j.minpro.2016.11.016
32. Deniz, V.: The effect of mill speed on kinetic breakage parameters of clinker and limestone. Cem. Concr. Res. **34**, 1365–1371 (2004). https://doi.org/10.1016/j.cemconres.2003.12.025
33. Deniz, V.: The effects of ball filling and ball diameter on kinetic breakage parameters of barite powder. Adv. Powder Technol. **23**, 640–646 (2012). https://doi.org/10.1016/j.apt.2011.07.006
34. Celik, M.S.: A comparison of dry and wet fine grinding of coals in a ball mill. Powder Technol. **55**, 1–9 (1988). https://doi.org/10.1016/0032-5910(88)80082-2
35. Bu, X., Ma, G., Peng, Y., Xie, G., Zhan, H., Liu, B.: Grinding kinetics of coal in wet ball-milling using the Taguchi method. Int. J. Coal Prep. Util. (2019). https://doi.org/10.1080/19392699.2019.1603147
36. Umucu, Y., Deniz, V., Cayirli, S.: A new model for comminution behavior of different coals in an impact crusher. Energy Sources, Part A Recover. Util. Environ. Eff. **36**, 1406–1413 (2014). https://doi.org/10.1080/15567036.2010.503232
37. Kotake, N., Kuboki, M., Kiya, S., Kanda, Y.: Influence of dry and wet grinding conditions on fineness and shape of particle size distribution of product in a ball mill. Adv. Powder Technol. **22**, 86–92 (2011). https://doi.org/10.1016/j.apt.2010.03.015
38. Klimpel, R.R.: Evaluating comminution efficiency from the point of view of downstream froth flotation. Miner. Metall. Process. **15**, 1–8 (1998). https://doi.org/10.1007/bf03403150
39. Koleini, S.M.J., Abdollahy, M., Soltani, F.: Wet and dry grinding methods effect on the flotation of taknar Cu-Zn sulphide ore using a mixed collector. In: 26th International Mineral Processing Congress, IMPC 2012: Innovative Processing for Sustainable Growth—Conference Proceedings, pp. 5113–5119 (2012)
40. Jung, H.J., Sohn, Y., Sung, H.G., Hyun, H.S., Shin, W.G.: Physicochemical properties of ball milled boron particles: dry vs. wet ball milling process. Powder Technol. **269**, 548–553 (2015). https://doi.org/10.1016/j.powtec.2014.03.058
41. Bu, X., Xie, G., Chen, Y., Ni, C.: The order of kinetic models in coal fines flotation. Int. J. Coal Prep. Util. **37**, 113–123 (2017). https://doi.org/10.1080/19392699.2016.1140150
42. Marwa, E.M.M., Meharg, A.A., Rice, C.M.: The effect of heating temperature on the properties of vermiculites from Tanzania with respect to potential agronomic applications. Appl. Clay Sci. **43**, 376–382 (2009). https://doi.org/10.1016/j.clay.2008.11.005
43. Derkowski, A., Franus, W., Beran, E., Czímerová, A.: Properties and potential applications of zeolitic materials produced from fly ash using simple method of synthesis. Powder Technol. **166**, 47–54 (2006). https://doi.org/10.1016/j.powtec.2006.05.004
44. Addai-Mensah, J., Ralston, J.: Investigation of the role of interfacial chemistry on particle interactions, sedimentation and electroosmotic dewatering of model kaolinite dispersions. In: Proceedings of the Powder Technology, vol. 160, pp. 35–39. Elsevier (2005)
45. Sondi, I., Lojen, S., Juračić, M., Prohić, E.: Mechanisms of land-sea interactions—the distribution of metals and sedimentary organic matter in sediments of a river-dominated Mediterranean karstic estuary. Estuar. Coast. Shelf Sci. **80**, 12–20 (2008). https://doi.org/10.1016/j.ecss.2008.07.001
46. Kotake, N., Kanda, Y., Koizumi, H., Kawaguchi, T.: A fundamental study of dry and wet grinding from the viewpoint of bending tests by drop weight method. Int. J. Soc. Mater. Eng. Resour. **11**, 1–4 (2003). https://doi.org/10.5188/ijsmer.11.1
47. Bu, X., Chen, Y., Ma, G., Sun, Y., Ni, C., Xie, G.: Wet and dry grinding of coal in a laboratory-scale ball mill: Particle-size distributions. Powder Technol. **359**, 305–313 (2020). https://doi.org/10.1016/j.powtec.2019.09.062
48. Tangsathitkulchai, C., Austin, L.G.: The effect of slurry density on breakage parameters of quartz, coal and copper ore in a laboratory ball mill. Powder Technol. **42**, 287–296 (1985). https://doi.org/10.1016/0032-5910(85)80068-1
49. Prziwara, P., Breitung-Faes, S., Kwade, A.: Impact of grinding aids on dry grinding performance, bulk properties and surface energy. Adv. Powder Technol. **29**, 416–425 (2018). https://doi.org/10.1016/j.apt.2017.11.029

50. Peukert, W., Schwarzer, H.C., Stenger, F.: Control of aggregation in production and handling of nanoparticles. In: Proceedings of the Chemical Engineering and Processing: Process Intensification, vol. 44, pp. 245–252. Elsevier (2005)
51. Knieke, C., Sommer, M., Peukert, W.: Identifying the apparent and true grinding limit. Powder Technol. **195**, 25–30 (2009). https://doi.org/10.1016/j.powtec.2009.05.007
52. Stražišar, J., Runovc, F.: Kinetics of comminution in micro- and sub-micrometer ranges. In: Comminution 1994, pp. 673–682. Elsevier (1996)
53. Zhang, Q., Kano, J., Saito, F.: Chapter 11 Fine grinding of materials in dry systems and mechanochemistry. Handb. Powder Technol. **12**, 509–528 (2007)
54. Boldyrev, V.V., Pavlov, S.V., Goldberg, E.L.: Interrelation between fine grinding and mechanical activation. Int. J. Miner. Process. **44–45**, 181–185 (1996). https://doi.org/10.1016/0301-7516(95)00028-3
55. Kendall, K.: The impossibility of comminuting small particles by compression. Nature **272**, 710–711 (1978). https://doi.org/10.1038/272710a0
56. Opoczky, L.: Fine grinding and agglomeration of silicates. Powder Technol. **17**, 1–7 (1977). https://doi.org/10.1016/0032-5910(77)85037-7
57. Kohobhange, S.P.K., Manoratne, C.H., Pitawala, H.M.T.G.A., Rajapakse, R.M.G.: The effect of prolonged milling time on comminution of quartz. Powder Technol. **330**, 266–274 (2018). https://doi.org/10.1016/j.powtec.2018.02.033
58. Palaniandy, S., Azizli, K.A.M., Hussin, H., Hashim, S.F.S.: Study on mechanochemical effect of silica for short grinding period. Int. J. Miner. Process. **82**, 195–202 (2007). https://doi.org/10.1016/j.minpro.2006.10.008
59. Hasegawa, M., Kimata, M., Shimane, M., Shoji, T., Tsuruta, M.: The effect of liquid additives on dry ultrafine grinding of quartz. Powder Technol. **114**, 145–151 (2001). https://doi.org/10.1016/S0032-5910(00)00290-4
60. Wahyudi, A., Nurasid, T., Rochani, S.: Preparation of nanoparticle silica from silica sand and quartzite by ultrafine grinding. In: Proceedings of International Conference on Chemical and Materials Engineering, pp. 1–7 (2012)
61. Palaniandy, S., Azizi Mohd Azizli, K., Hussin, H., Fuad Saiyid Hashim, S.: Mechanochemistry of silica on jet milling. J. Mater. Process. Technol. **205**, 119–127 (2008). https://doi.org/10.1016/j.jmatprotec.2007.11.086
62. Wills' Mineral Processing Technology. Elsevier (2016)
63. Tripathi, A., Sankrityayanand, U., Gupta, V.K.: Effect of particle size distribution on grinding kinetics in dry and wet ball milling operations. In: 26th International Mineral Processing Congress, IMPC 2012: Innovative Processing for Sustainable Growth—Conference Proceedings, pp. 5500–5507 (2012)
64. Varas, R.A.C., Valenzuela, M.A.: Empirical determination of the effect of lifter wear in mill power for dry grinding. IEEE Trans. Ind. Appl. **53**, 2621–2627 (2017). https://doi.org/10.1109/TIA.2017.2661844
65. Opazo, B.C., Anibal Valenzuela, M.: Experimental evaluation of power requirements for wet grinding and its comparison to dry grinding. IEEE Trans. Ind. Appl. **54**, 3953–3960 (2018). https://doi.org/10.1109/TIA.2018.2821100
66. Soleymani, M.M., Fooladi, M., Rezaeizadeh, M.: Experimental investigation of the power draw of tumbling mills in wet grinding. Proc. Inst. Mech. Eng. Part C J. Mech. Eng. Sci. **230**, 2709–2719 (2016). https://doi.org/10.1177/0954406215598801
67. Powell, M.S., McBride, A.T.: A three-dimensional analysis of media motion and grinding regions in mills. In: Proceedings of the Minerals Engineering, vol. 17, pp. 1099–1109. Pergamon (2004)
68. Shi, F.N., Napier-Munn, T.J.: Estimation of shear rates inside a ball mill. Int. J. Miner. Process. **57**, 167–183 (1999). https://doi.org/10.1016/S0301-7516(99)00016-2
69. Shahbazi, B., Chehreh Chelgani, S., Matin, S.S.: Prediction of froth flotation responses based on various conditioning parameters by Random Forest method. Colloids Surfaces A Physicochem. Eng. Asp. **529**, 936–941 (2017). https://doi.org/10.1016/j.colsurfa.2017.07.013

70. Leistner, T., Embrechts, M., Leißner, T., Chehreh Chelgani, S., Osbahr, I., Möckel, R., Peuker, U.A., Rudolph, M.: A study of the reprocessing of fine and ultrafine cassiterite from gravity tailing residues by using various flotation techniques. Miner. Eng. **96–97**, 94–98 (2016). https://doi.org/10.1016/j.mineng.2016.06.020
71. Seke, M.D., Pistorius, P.C.: Effect of cuprous cyanide, dry and wet milling on the selective flotation of galena and sphalerite. Miner. Eng. **19**, 1–11 (2006). https://doi.org/10.1016/j.mineng.2005.03.005
72. Liu, J., Long, H., Corin, K.C., O'Connor, C.T.: A study of the effect of grinding environment on the flotation of two copper sulphide ores. Miner. Eng. **122**, 339–345 (2018). https://doi.org/10.1016/j.mineng.2018.03.031
73. Learmont, M.E., Iwasaki, I.: Effect of grinding media on galena flotation. Miner. Metall. Process. **1**, 136–143 (1984). https://doi.org/10.1007/bf03402566
74. Lowrison, G.C.: Crushing and grinding: the size reduction of solid materials, pp. 1–286 (1974)
75. Suzuki, A., Tanaka, T.: Crushing efficiency in relation to some operational variables and material constants. Ind. Eng. Chem. Process Des. Dev. **7**, 161–166 (1968). https://doi.org/10.1021/i260026a001
76. Lin, I.J., Somasundaran, P.: Alterations in properties of samples during their preparation by grinding. Powder Technol. **6**, 171–179 (1972). https://doi.org/10.1016/0032-5910(72)80074-3
77. Austin, L.G., Klimpel, R.R., Luckie, P.T.: Process engineering of size reduction: ball milling. Process Eng Size Reduct, Ball Milling (1984)
78. Nie, Y., Chen, J., Wang, Q., Zhang, C., Shi, C., Zhao, J.: Use of dry grinding process to increase the leaching of gold from a roasted concentrate containing hematite in the thiosulfate system. Hydrometallurgy **201**, 105582 (2021). https://doi.org/10.1016/j.hydromet.2021.105582
79. Feng, D., van Deventer, J.S.J.: The effect of iron contaminants on thiosulphate leaching of gold. Miner. Eng. **23**, 399–406 (2010). https://doi.org/10.1016/j.mineng.2009.11.016
80. Feng, D., van Deventer, J.S.J.: Effect of hematite on thiosulphate leaching of gold. Int. J. Miner. Process. **82**, 138–147 (2007). https://doi.org/10.1016/j.minpro.2006.09.003
81. Rácz, Á.: Research and development of the grinding process for the production of ultrafine materials (2014)
82. Saramak, D., Kleiv, R.A.: The effect of feed moisture on the comminution efficiency of HPGR circuits. Miner. Eng. **43–44**, 105–111 (2013). https://doi.org/10.1016/j.mineng.2012.09.014
83. Rashidi, S., Rajamani, R.K., Fuerstenau, D.W.: A review of the modeling of high pressure grinding rolls. KONA Powder Part. J. **2017**, 125–140 (2017)
84. Daniel, M.: Michael Energy efficient mineral liberation using HPGR technology (2007)
85. Mclvor, R.E.: High pressure grinding rolls—a review, pp. 15–39 (2006)
86. Altun, O., Benzer, H., Dundar, H., Aydogan, N.A.: Comparison of open and closed circuit HPGR application on dry grinding circuit performance. Miner. Eng. **24**, 267–275 (2011). https://doi.org/10.1016/j.mineng.2010.08.024
87. Rosario, P., Hall, R.: Analyses of the Total Required Energy for Comminution of Hard Ores in Sag Mill and Hpgr Circuits, pp. 129–138 (2008)
88. Prziwara, P., Breitung-Faes, S., Kwade, A.: Comparative study of the grinding aid effects for dry fine grinding of different materials. Miner. Eng. **144**, 106030 (2019). https://doi.org/10.1016/j.mineng.2019.106030
89. Gokcen, H.S., Cayirli, S., Ucbas, Y., Kayaci, K.: The effect of grinding aids on dry micro fine grinding of feldspar. Int. J. Miner. Process. **136**, 42–44 (2015). https://doi.org/10.1016/J.MINPRO.2014.10.001
90. Chipakwe, V., Semsari, P., Karlkvist, T., Rosenkranz, J., Chelgani, S.C.: A critical review on the mechanisms of chemical additives used in grinding and their effects on the downstream processes. J. Mater. Res. Technol. **9**, 8148–8162 (2020). https://doi.org/10.1016/j.jmrt.2020.05.080
91. Prziwara, P., Hamilton, L.D., Breitung-Faes, S., Kwade, A.: Impact of grinding aids and process parameters on dry stirred media milling. Powder Technol. **335**, 114–123 (2018). https://doi.org/10.1016/j.powtec.2018.05.021

92. Katsioti, M., Tsakiridis, P.E., Giannatos, P., Tsibouki, Z., Marinos, J.: Characterization of various cement grinding aids and their impact on grindability and cement performance. Constr. Build. Mater. **23**, 1954–1959 (2009). https://doi.org/10.1016/J.CONBUILDMAT.2008.09.003
93. Assaad, J.J.: Quantifying the effect of clinker grinding aids under laboratory conditions. Miner. Eng. **81**, 40–51 (2015). https://doi.org/10.1016/J.MINENG.2015.07.008
94. Toprak, N.A., Benzer, A.H., Karahan, C.E., Zencirci, E.S.: Effects of grinding aid dosage on circuit performance and cement fineness. Constr. Build. Mater. **265**, 120707 (2020). https://doi.org/10.1016/J.CONBUILDMAT.2020.120707
95. Zheng, J., Harris, C.C., Somasundaran, P.: The effect of additives on stirred media milling of limestone. Powder Technol. **91**, 173–179 (1997)
96. Mishra, R.K., Flatt, R.J., Heinz, H.: Force field for tricalcium silicate and insight into nanoscale properties: cleavage, initial hydration, and adsorption of organic molecules. J. Phys. Chem. C **117**, 10417–10432 (2013). https://doi.org/10.1021/JP312815G
97. Weibel, M., Mishra, R.K.: Comprehensive understanding of grinding aids. ZKG Int. **67**, 28–39 (2014)
98. Sottili, L., D.P.: Effect of grinding admixtures in the cement industry, Part 1. ZKG Int. **10** (2000)
99. Paramasivam, R., Vedaraman, R.: Studies in additive grinding of minerals. Adv. Powder Technol. **3**, 31–37 (1992). https://doi.org/10.1016/S0921-8831(08)60686-X
100. Shi, F., Morrison, R., Cervellin, A., Burns, F., Musa, F.: Comparison of energy efficiency between ball mills and stirred mills in coarse grinding. Miner. Eng. **22**, 673–680 (2009). https://doi.org/10.1016/J.MINENG.2008.12.002
101. Kwade, A., Schwedes, J.: Chapter 6 Wet grinding in stirred media mills. Handb. Powder Technol. **12**, 251–382 (2007). https://doi.org/10.1016/S0167-3785(07)12009-1
102. Oliveira, A.L.R., Rodriguez, V.A., de Carvalho, R.M., Powell, M.S., Tavares, L.M.: Mechanistic modeling and simulation of a batch vertical stirred mill. Miner. Eng. **156**, 106487 (2020). https://doi.org/10.1016/J.MINENG.2020.106487
103. Altun, O., Benzer, H., Enderle, U.: Effects of operating parameters on the efficiency of dry stirred milling. Miner. Eng. **43–44**, 58–66 (2013). https://doi.org/10.1016/j.mineng.2012.08.003
104. Yang, Y., Rowson, N.A., Tamblyn, R., Ingram, A.: Effect of operating parameters on fine particle grinding in a vertically stirred media mill. Sep. Sci. Technol. **52**, 1143–1152 (2017). https://doi.org/10.1080/01496395.2016.1276931
105. Wang, Y., Forssberg, E.: Product size distribution in stirred media mills. Miner. Eng. **13**, 459–465 (2000). https://doi.org/10.1016/S0892-6875(00)00025-X
106. Theuerkauf, J., Schwedes, J.: Theoretical and experimental investigation on particle and fluid motion in stirred media mills. Powder Technol. **105**, 406–412 (1999). https://doi.org/10.1016/S0032-5910(99)00165-5
107. Radziszewski, P.: Assessing the stirred mill design space. Miner. Eng. **41**, 9–16 (2013). https://doi.org/10.1016/J.MINENG.2012.10.012
108. Ohenoja, K., Illikainen, M., Niinimäki, J.: Effect of operational parameters and stress energies on the particle size distribution of TiO2 pigment in stirred media milling. Powder Technol. **234**, 91–96 (2013). https://doi.org/10.1016/J.POWTEC.2012.09.038
109. Blecher, L., Kwade, A., Schwedes, J.: Motion and stress intensity of grinding beads in a stirred media mill. Part 1: energy density distribution and motion of single grinding beads. Powder Technol. **86**, 59–68 (1996). https://doi.org/10.1016/0032-5910(95)03038-7
110. Xiao, X., Zhang, G., Feng, Q., Xiao, S., Huang, L., Zhao, X., Li, Z.: The liberation effect of magnetite fine ground by vertical stirred mill and ball mill. Miner. Eng. **34**, 63–69 (2012). https://doi.org/10.1016/J.MINENG.2012.04.004
111. Zhang, W., Yang, J., Wu, X., Hu, Y., Yu, W., Wang, J., Dong, J., Li, M., Liang, S., Hu, J., et al.: A critical review on secondary lead recycling technology and its prospect. Renew. Sustain. Energy Rev. **61**, 108–122 (2016). https://doi.org/10.1016/j.rser.2016.03.046

112. Kwade, A.: Mill selection and process optimization using a physical grinding model. Int. J. Miner. Process. **74**, S93–S101 (2004). https://doi.org/10.1016/J.MINPRO.2004.07.027
113. Jankovic, A., Valery, W., Rosa, D.L.: Fine grinding in the Australian mining industry. Metso Miner. Process … 1–11 (2008)
114. Wang, Y., Forssberg, E., Sachweh, J.: Dry fine comminution in a stirred media mill—MaxxMill®. Int. J. Miner. Process. **74**, S65–S74 (2004). https://doi.org/10.1016/J.MINPRO.2004.07.010
115. Pilevneli, C.C., Kizgut, S., Toroglu, I., Çuhadaroglu, D., Yigit, E.: Open and closed circuit dry grinding of cement mill rejects in a pilot scale vertical stirred mill. Powder Technol. **139**, 165–174 (2004). https://doi.org/10.1016/J.POWTEC.2003.12.002
116. Jankovic, A.: Variables affecting the fine grinding of minerals using stirred mills. Miner. Eng. **16**, 337–345 (2003). https://doi.org/10.1016/S0892-6875(03)00007-4
117. Bel Fadhel, H., Frances, C.: Wet batch grinding of alumina hydrate in a stirred bead mill. Powder Technol. **119**, 257–268 (2001). https://doi.org/10.1016/S0032-5910(01)00266-2
118. Altun, O., Benzer, H., Karahan, E., Zencirci, S., Toprak, A.: The impacts of dry stirred milling application on quality and production rate of the cement grinding circuits. Miner. Eng. **155**, 106478 (2020). https://doi.org/10.1016/J.MINENG.2020.106478
119. Mucsi, G., Rácz, Á., Mádai, V.: Mechanical activation of cement in stirred media mill. Powder Technol. **235**, 163–172 (2013). https://doi.org/10.1016/J.POWTEC.2012.10.005
120. Altun, O.: Energy and cement quality optimization of a cement grinding circuit. Adv. Powder Technol. **29**, 1713–1723 (2018). https://doi.org/10.1016/J.APT.2018.04.006
121. Gerl, S., Sachweh, J.: Plant concepts for ultrafine dry grinding with the agitated media mill MaxxMill®. Miner. Eng. **20**, 327–333 (2007). https://doi.org/10.1016/J.MINENG.2006.09.007

Chapter 2
Magnetic Separation

2.1 Introduction

The differences in magnetic susceptibility can be utilized to separate a valuable mineral from its gangue through the magnetic separation method. Similar to other materials, minerals are generally classified into three main categories, namely diamagnetic, paramagnetic, and ferromagnetic. Among these subgroups, ferromagnetic minerals possess the highest magnetic susceptibilities; therefore, the efficiency of magnetic separation for them is higher than paramagnetic and diamagnetic minerals [1, 2]. When minerals are exposed to a magnetic field, various particles behave differently depending on the group they belong to [3]. Diamagnetic minerals are excreted to points that the intensity of the magnetic field is smaller than the first location. During magnetic separation, they do not experience a magnetic attraction force. That is why diamagnetic materials are also known as non-magnetic; however, this is not strictly accurate. On the other hand, paramagnetic minerals are attracted by magnetic fields and move along the lines of magnetic force to points of greater field intensity. Ferromagnetic materials can be classified as a special type of paramagnetic materials which is highly susceptible to magnetic forces and might become a permanent magnet if exposed to a magnetic field. For this reason, in a magnetic field, ferromagnetic particles quickly become adapted to the lines of the field [1, 4, 5]. Figure 2.1 shows the magnetization curves of various classes of materials [6]. Several minerals belonging to different groups are also listed in Table 2.1 [7, 8]. There is a strong correlation between minerals' magnetic susceptibility and their chemical compositions [9]. Generally, besides reducing water consumption, the use of dry magnetic separation methods also has other benefits. In dry magnetic separators, usually, more precise separation is achievable. Additionally, dry magnets are generally more controllable than wet ones. The separation process is naturally easier without contesting the drag forces created by water. Furthermore, dry mode, more specifically when using rare-earth magnets, is lower in both capital and maintenance costs [10].

S. C. Chelgani and A. Asimi Neisiani, *Dry Mineral Processing*,
https://doi.org/10.1007/978-3-030-93750-8_2

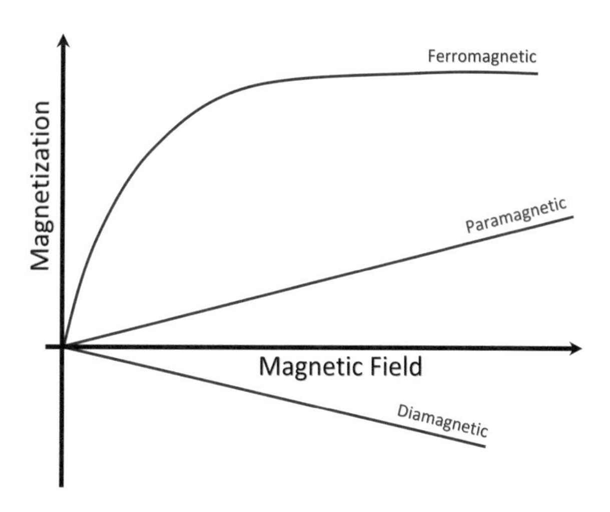

Fig. 2.1 Magnetization curves of various materials [6]

In mineral processing, magnetic separation is usually employed for three purposes:

- Ejection of iron and other similar materials present in the p stream to protect equipment against wear or damage,
- Separation of valuable magnetic minerals,
- Removal of magnetic impurities [11, 12].

As a step of discarding tailings before main processes, magnetic separators can considerably reduce process costs. Because by eliminating a significant portion of the tailings, magnetic separators can reduce the costs of screening, grinding, and dewatering. Additionally, magnetic separators can obtain coarse-grain marketable tailings, which decreases the cost of processing units [11, 12]. Magnetic-based separators can be categorized from different perspectives (Fig. 2.2) [13]. In this chapter, the most important dry magnetic separators are mentioned. Their advantages and weaknesses have also been discussed, along with their applications in mineral processing.

2.2 Dry Low-Intensity Magnetic Separators (DLIMS)

In General, large-sized ferromagnetic and paramagnetic minerals with high magnetic susceptibility are mainly separated by low-intensity separators. These separators can be applied in both wet and dry modes. On the other hand, high-intensity magnetic separators can be utilized to treat weakly magnetic materials, whether fine or coarse, in dry or wet mode. In the case of fine, feebly magnetic minerals, the best choice for magnetic separation is high-gradient magnetic separators [13].

Table 2.1 Minerals and their magnetic response [7, 8]

Mineral name	Chemical formula	Magnetic response
Magnetite	Fe_3O_4	Ferromagnetic
Franklinite	$(Zn,Mn)Fe_2O_4$	Ferromagnetic
Pyrrhotite	$Fe_{(1-x)}S$	Ferromagnetic
Amphibole	$(Fe,Mg,Ca)_xSiO_3$	Paramagnetic
Geothite	$FeO(OH)$	Paramagnetic
Chromite	$(Fe,Mg)(Cr,Al)_2O_4$	Paramagnetic
Hematite	Fe_2O_3	Paramagnetic
Ilmenite	$FeTiO_3$	Paramagnetic
Monazite	$(Ce,La,Y,Th)PO_4$	Paramagnetic
Niccolite	$NiAs$	Paramagnetic
Olivine	$(Mg,Fe)_2[SiO_4]$	Paramagnetic
Siderite	$FeCO_3$	Paramagnetic
Uraninite	UO_2	Paramagnetic
Wolframite	$(Fe,Mn)WO_4$	Paramagnetic
Xenotime	YPO_4	Paramagnetic
Orpiment	As_2S_3	Diamagnetic
Muscovite	$KAl_2[AlSi_3O_{10}][F,OH]_2$	Diamagnetic
Molybdenite	MoS_2	Diamagnetic
Malachite	$Cu_2CO_3(OH)_2$	Diamagnetic
Kaolinite	$Al_2Si_2O_5(OH)_4$	Diamagnetic
Calcite	$CaCO_3$	Diamagnetic
Chalcocite	Cu_2S	Diamagnetic
Corumdum	Al_2O_3	Diamagnetic
Cuprite	Cu_2O	Diamagnetic
Diamond	C	Diamagnetic
Dolomite	$CaMg(CO_3)_2$	Diamagnetic
Feldspar group	$(K,Na,Ca..)x(Al,Si)_3O_8$	Diamagnetic
Fluorite	CaF_2	Diamagnetic
Galena	PbS	Diamagnetic
Gold	Au	Diamagnetic
Graphite	C	Diamagnetic
Gypsum	$CaSO_4 \cdot 2H_2O$	Diamagnetic
Quartz	SiO_2	Diamagnetic
Rutile	TiO_2	Diamagnetic
Smithsonite	$ZnCO_3$	Diamagnetic
Talc	$Mg_3Si_4O_{10}(OH)_2$	Diamagnetic
Topaz	$Al_2SiO_4(F,OH)_2$	Diamagnetic
Zincite	ZnO	Diamagnetic
Zircon	$ZrSiO_4$	Diamagnetic

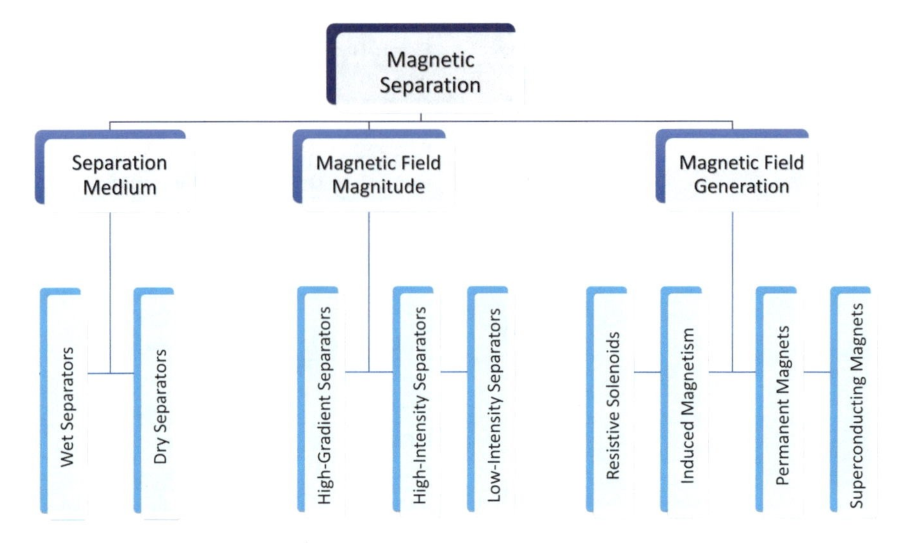

Fig. 2.2 Classification of magnetic separators from different perspectives [13]

Dry low-intensity magnetic separators (DLIMS) are generally employed to:

- upgrade iron ores containing relatively large particles and strongly magnetic minerals
- separate iron pieces from steel mill slags and blast furnaces
- remove impurities with strongly magnetic characteristics
- remove tramp iron [1, 14].

For removing unfavorable iron pieces, a DLIMS called tramp iron magnetic separator has been designed and used in mineral processing plants to keep safe various equipment, such as material handling systems, crushers, and mills. This type of separator is designed for materials that are dry or contain only surface moisture. Tramp iron magnetic separator is able to remove large particles up to 2 m in size successfully. When selecting the required protective magnet, parameters such as the shape and size of iron pieces as well as the material handling system have to be taken into consideration [1, 15].

2.2.1 *Magnetic Pulleys*

A magnetic head pulley is a separator by which tramp iron can be easily eliminated from materials on a conveyor belt. This separator is designed with either electromagnetic construction or a permanent magnet. These comparatively inexpensive separators are readily installed and can achieve continuous tramp iron removal. The pully of the conveyor belt can be replaced by a revolving magnetic pulley, leading to converting the conveyor belt into a magnetic separator (Fig. 2.3) [16, 17].

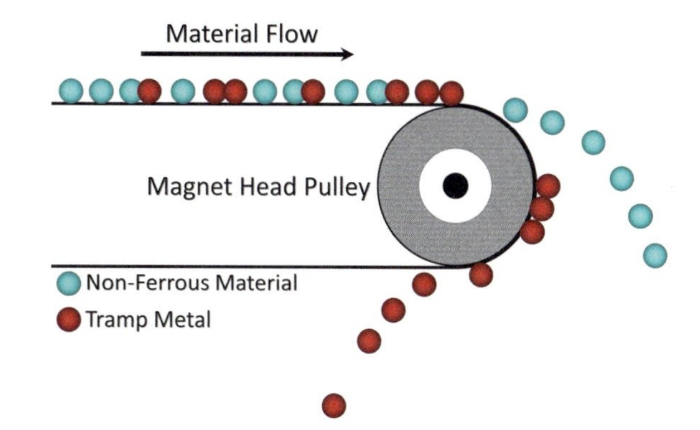

Fig. 2.3 Tramp iron removal using magnetic pulley [17]

Magnetic particles are carried by the conveyor belt and remain attached to it until they pass through the magnetic field region. On the other hand, non-magnetic particles, which are not affected by magnetic force, are discharged over the pulley. Permanent magnetic pulleys, which are usually made of strontium ferrite ($SrO{\cdot}6Fe_2O_3$), are designed in two ways:

- axial poles design in which the polarity alternates along the circumference
- radial poles design in which the polarity changes across the width of the belt.

The former is more suitable for finely divided particles because of the uniformity of the magnetic field across the conveyor belt width. However, radial poles design is more advantageous to remove elongated-shaped and large-sized materials. When the magnetic field should be controlled, electromagnetic pulleys are used [18]. When selecting the magnetic pulley, it should be considered that the width of the separator must be matched with the conveyor belt width. The magnetic pulleys are constructed in a range of diameters from 50 to 125 cm. Inclined belts have also been designed to provide further areas of contact with the pulley's magnetic field and, therefore, can increase the capacity. Moreover, some magnetic pulleys are specially designed to concentrate magnetite and other ferromagnetic minerals [17]. There are specially designed pulleys that are industrially utilized to upgrade ores containing ferromagnetic minerals such as magnetite. For making an efficient process, the feed should be divided into different size fractions through screening, and subsequently, each fraction can be treated on a certain pulley. Additionally, it should be noted that magnetic pulleys are not suitable to separate the particles of −6 mm diameter. Table 2.2 illustrates recommended pulley diameters for treatment of various particle sizes [16, 17]

Table 2.2 Recommended diameter of magnetic pulley [17]

Feed size (mm)	Pulley diameter (mm)	Throughput ($t\,h^{-1}\,m^{-1}$)
50–100	750–900	250–400
25–50	450–600	165–250
6–25	300–450	75–135

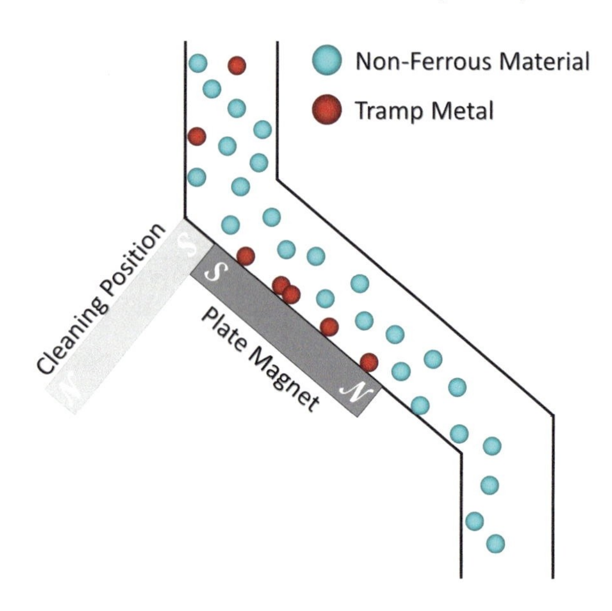

Fig. 2.4 Schematic diagram of separation using plate magnet [16]

2.2.2 *Magnetic Plate*

In materials with a small amount of tramp iron, automatic removal of iron pieces is unnecessary, and the magnetic plate separators can be effectively used. These types of DLIMS can be incorporated in ducts and chutes to eliminate tramp iron. When using a plate magnet in a chute (Fig. 2.4), tramp iron is attracted and trapped by magnetic force, whereas the remaining materials (non-magnetic portion) pass in front of the magnetic plate without being absorbed. The magnetized plate must be periodically inspected and cleaned because its efficiency decreases if tramp iron particles accumulate. The plate magnet separators are usually manufactured with permanent magnets made of rare earth elements or ferrite. Depending on the shape of materials to be eliminated and their type, the effective depth of the magnetic field can vary; however, it can be generally as high as 200 mm. Also, the chute angle should not exceed 45° [16, 17].

2.2.3 *Magnetic Grate*

The magnetic grate, also known as hopper magnet (Fig. 2.5), comprises several tubes in which magnets are embedded and usually mounted at the discharge section of a hopper. Magnetic tube raws can provide several streams of finely divided particles and then eliminate magnetic materials. Analogous to plate magnets, this type of separator can be effective when the amount of iron pieces in the input material is low. Magnetic grates are recommended for granular, free-flowing particles, specifically when finer

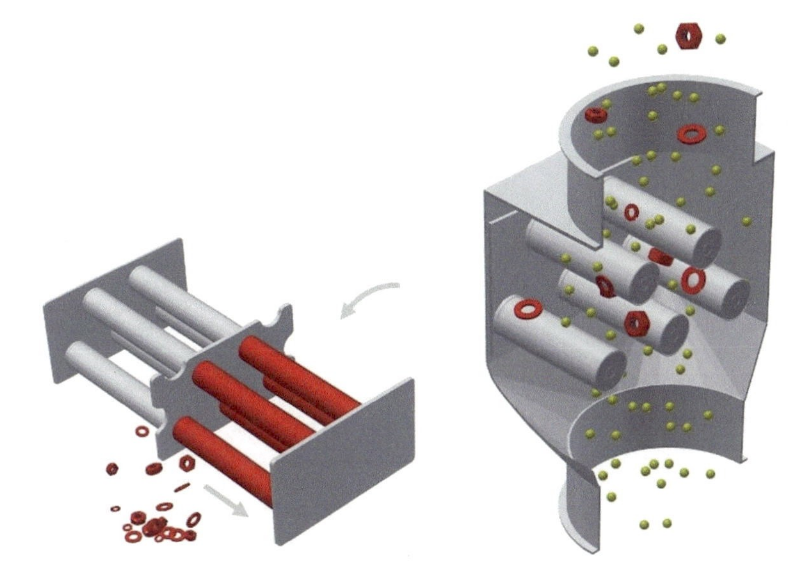

Fig. 2.5 Magnetic grate separator [19]

than 12 mm. Different materials can be used to make this type of separator depending on the operating temperature. NdFeB magnets are suitable for temperatures up to 90 °C, whereas ferrite-based magnets are chosen for higher operating temperatures (up to 150 °C). Additionally, Alnico (A-Fe alloy) magnetic grates can be employed when the operating temperature is up to 350 °C [17].

2.2.4 *Suspended Magnets*

Suspended magnets have been frequently utilized to protect the equipment against damage caused by metal pieces and enhance the quality of materials by eliminating magnetic particles. This type of separator, which is designed in rectangular or circular construction, is one of the best options, especially when the speed of the conveyor belt is high and a small amount of tramp iron is present in the material stream. Suspended magnets can be installed at several locations of material transport systems, such as any desired point in conveyor belt lengthways, above chutes or launders, the discharge part of screens and feeders. However, the conveyor belt's angles are the ideal points for this separator. In this way, the iron pieces are exposed to the separator's face, and therefore, the tramp iron can be removed more easily [20]. Simply suspended magnets have to be manually cleaned of iron accumulated and maintained at regular intervals. In cases where the material stream contains a significant amount of tramp iron or the magnet is difficult to clean, a self-cleaning mechanism is recommended. This mechanism causes tramp iron to be removed continuously and automatically from

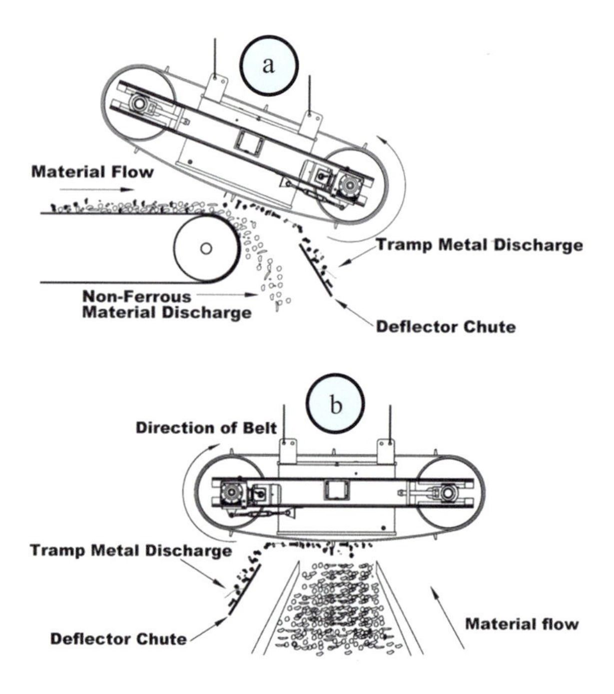

Fig. 2.6 Inline (**a**) and cross-belt (**b**) position of suspended magnetic separator [17]

the material stream. Separators equipped with self-cleaning magnets are designed cross-belt or in-line (Fig. 2.6) [17]. In the cross-belt type, the suspended magnet belt's movement direction is perpendicular to the main conveyor belt's movement. Cross-belt magnets are mainly stronger and larger than in-line type because tramp iron needs to be turned by 90° from the movement direction on the conveyor belt [17].

2.2.5 *Magnetic Drum*

The most extensively magnetic-based separators in mineral processing engineering are magnetic drums. Drum magnets are mainly manufactured based on permanent magnets. However, electromagnetic drums are occasionally used in industrial applications. These separators can cover a wide range of particles from several centimeters down to several micrometers and operate in both wet and dry mode. In conventional magnetic drums, ferrite magnets are used. However, powerful high-intensity magnetic drums have been manufactured with rare-earth permanent magnets. The basic design of all low-intensity magnetic drums is principally the same. A set of permanent magnets are embedded inside a non-magnetic revolving drum. Based on the arrangement of magnets inside the drum, the separator can be divided into two subgroups, namely radial poles and axial poles types (Fig. 2.7) [1].

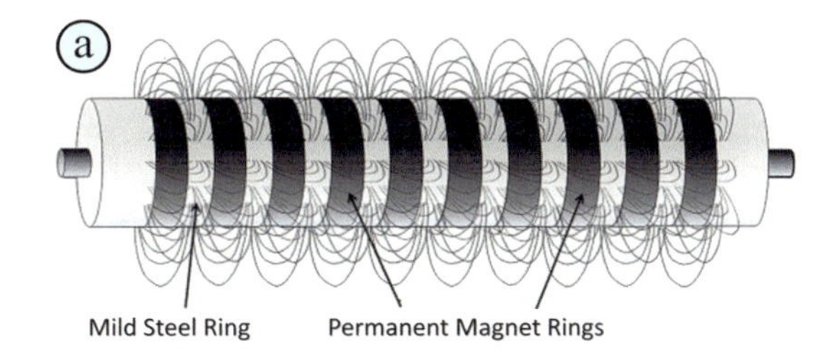

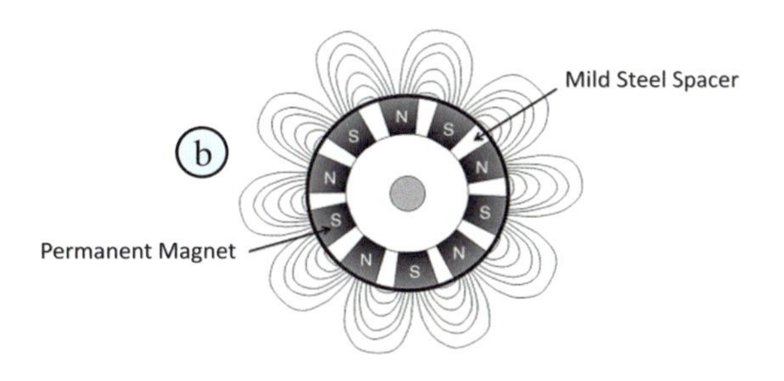

Fig. 2.7 Radial pole (**a**) and axial pole (**b**) drum magnets [21]

Each of these categories is used for a specific purpose. In the radial-arranged type, the polarity varies across the drum width and is uniform along the drum circumference. This arrangement has caused the machine to be suitable for eliminating elongated tramp iron and the concentration of coarse particles when high recovery is required. In the axial arrangement, the polarity varies along the drum circumference and is the same across the drum width. The axial arrangement can provide an agitating action, enabling the release of non-magnetic particles entrained by magnetic materials. This separator is usually utilized when a high-quality concentrate is required, even if sometimes this purpose reduces recovery. The development of new permanent magnets has led to enhanced magnetic induction available on the drum surface and at a suitable operating distance from the drum. Conventional magnetic drum separators, made of ferrite, can generate up to around 0.22 T on the surface of the drum and approximately 0.10 T at the distance of 50 mm. In new NdFeB-based drum magnets, however, this value can reach around 1 T on the surface of the drum. That is why using rare-earth separators, iron-bearing minerals, and stainless steel can be separated. Dry low-intensity drum magnetic separators are designed with different widths and diameters. In general, the diameters of ferrite magnetic drums can be from 60 to 150 cm, whereas in the case of rare-earth types, the diameters are generally smaller (38–100 cm). The most common use of drums is eliminating tramp iron, specifically when suspended and pulley magnets are not suitable. Drums can be used with top feed and bottom feed conditions (Fig. 2.8). Top feed drums can provide higher magnetic particle removal; therefore, they are preferable for loads containing a comparatively lower amount of magnetic materials [16, 17].

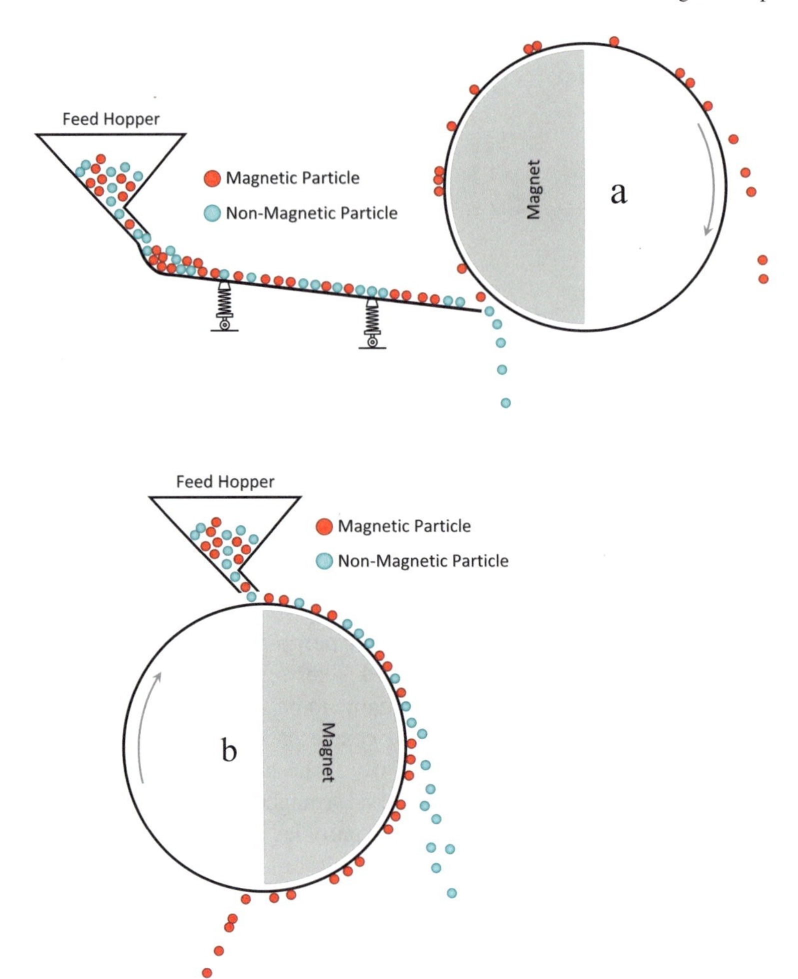

Fig. 2.8 Bottom feed (**a**) and top feed (**b**) magnetic drum separators [17]

2.3 Dry High-Intensity Magnetic Separators (DHIMS)

DHIMS can be divided into several permanent magnetic and electromagnetic separators (Fig. 2.9). A dry high-intensity magnetic separator is a waterless separation machine introduced to separate weakly magnetic particles that DLIMS cannot process. This technology employs a greater magnetic field strength provided by an induced or permanent magnet to separate the materials based on their magnetic susceptibility [20, 21]. Dry high-intensity magnetic separators can be used to separate paramagnetic particles in placers, eliminate magnetic impurities from glasses and refractory powders, and separate valuable minerals from various waste streams [11].

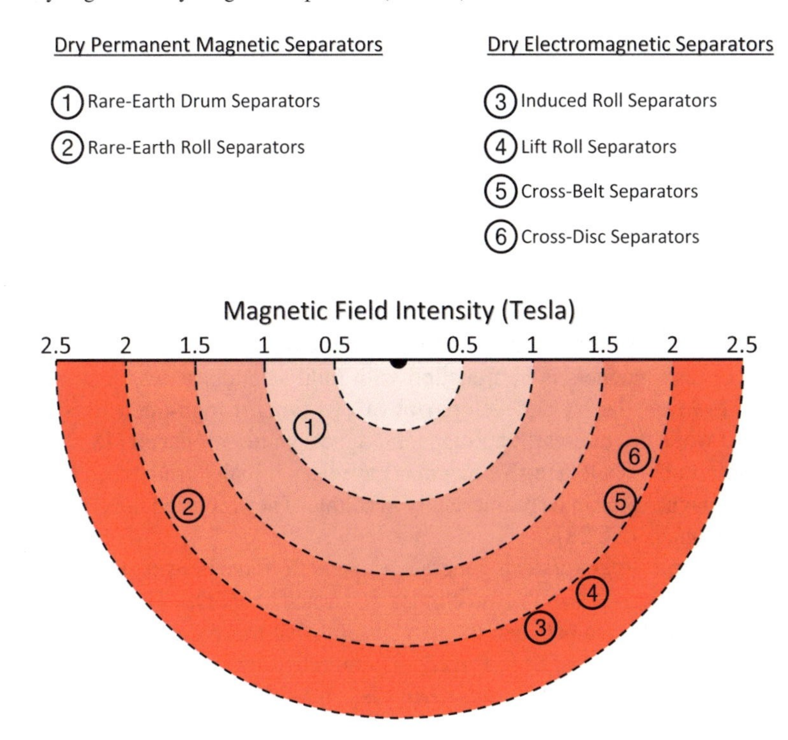

Fig. 2.9 Dry high-intensity magnetic separators [13]

2.3.1 Rare-Earth Drum Separator

These days, since rare-earth permanent magnets are becoming more available and affordable, they lead to the development and manufacture of powerful DHIMS and high-intensity magnetic drums [22]. Eriez, a magnetic separator manufacturer, was probably the first company that replaced standard ferrite-based drum magnets with permanent neodymium-iron-boron (NdFeB) magnets. Almost all rare-earth drum separators (REDS) suppliers follow Eriez's design philosophy [23]. However, the Permos separator, which uses small magnet blocks in its structure, was developed based on a fairly different mechanism in comparison to conventional REDS [24, 25]. It was claimed that this improved REDS needs less magnetic material than Eriez drum magnets to reach the magnetic induction of 0.7 T on the surface of the separator [25].

One of the features of REDS is the eddy current generation during the revolving of the drum's shell. This issue can be a major challenge since permanent NdFeB magnets can tolerate the limited temperature. The magnitude of the generated eddy currents depends on various parameters, such as the revolving speed of the drum, the number of magnetic poles, the magnetic induction, the conductivity of the shell material, and the shell thickness. It should be noted that the amount of heat generated is proportional to the required drive power to overcome the braking effect of the eddy currents. Sometimes factors such as the high temperature of the feed material intensify the heat generation in the NdFeB-base drum magnet. In such cases, due to the limited tolerance of this type of magnet to temperature, other separators with a greater temperature tolerance have to be utilized [26, 27].

2.3.2 Rare-Earth Roll Separator

The idea of the manufacture of rare-earth roll separators (RERS), comprising permanent ring or disk magnets in conjunction with mild steel discs, was first proposed in 1968. However, the rapid development of this type of high-intensity separator accelerated when the rare-earth permanent magnets were introduced [13, 26]. Even though samarium–cobalt (Sm-Co) magnets have been initially used, when (NdFe-B) permanent magnets became commercially available. They replaced the conventional rare-earth magnets [27, 28].

Rare-earth roll separators can generate a magnetic force density, also known as force index, much more than the force index generated by REDS. The main reason is the geometric difference between the two systems that make the magnetic gradient different. Furthermore, magnetic force can penetrate to greater depths, resulting in a more efficient separation of coarse particles. It is also stated that RERSs are a more powerful separation system than REDSs and can provide greater flexibility and selectivity [27]. The roll magnet (Fig. 2.10) is covered by a belt, supported by a non-magnetic roll because of the easy separation of magnetic materials. The thin belt is responsible for exposing the particles to the magnet. The magnetic field attracts magnetic particles, whereas non-magnetic materials are repelled because of magnetic, gravitational, centrifugal, and other forces. The thickness of the belt should be about 0.2 mm, permitting the particles to be close to the surface of the magnet.

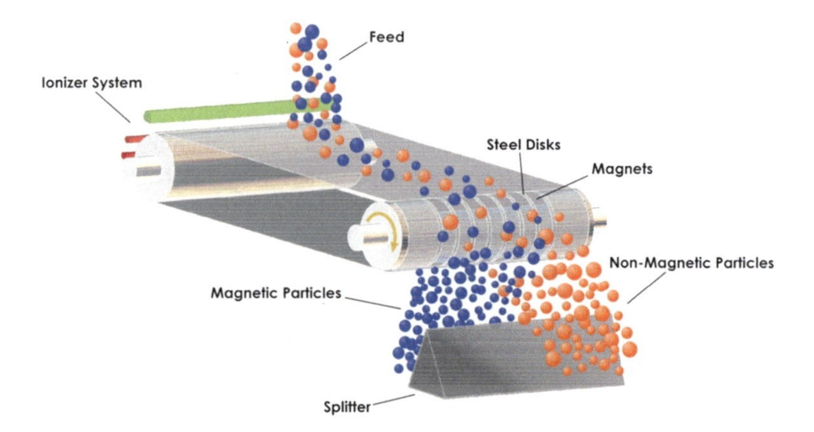

Fig. 2.10 Schematic diagram of a rare-earth roll separator [29]

Finally, A movable material splitter divide the magnetic and non-magnetic particle fractions into collection bins [1, 29].

The use of RERS increases separation efficiency without heat generation during separation. Similar to any other device, RERS has advantages and disadvantages. Lower energy consumption, smaller mass, and lower required space for installation are the most important benefits of RERS compared to REDS. RERS is an easily adjustable separator that can be adjusted according to the required product and feed variations. The drawbacks of this machine are its relatively high belt wear and the impossibility of controlling the magnetic field [26].

Since both RED and RER separators are common in processing units, when selecting them, a comparison of their performance for a specific feed is usually made. For example, the results of an investigation on the effect of feed rate for a garnet-rich heavy mineral sample (Fig. 2.11) indicated that RERS is very sensitive to feed rate; however, this parameter has not a significant effect on the REDS performance. As can be observed, using REDS, the garnet recovery decrease approximately 13% when the feed rate increases from 2.6 to 7.8 t/h. However, in the case of REDS, the reduction in recovery is marginal under the same conditions [30].

The capacity of RERS mainly depends on the roll diameter, which is designed and selected based on the particle size of the feed [31]. When using RERS, one of the parameters determining the cost of the separation process and affecting its flexibility is belt maintenance. The use of several tensioning systems and roll support structures has made it possible to replace the belt in less than 5 min [32]. Regarding RERS, parameters affecting the separation process can be divided into three groups:

- design variables
- process variables
- feed characteristics (Fig. 2.12) [13].

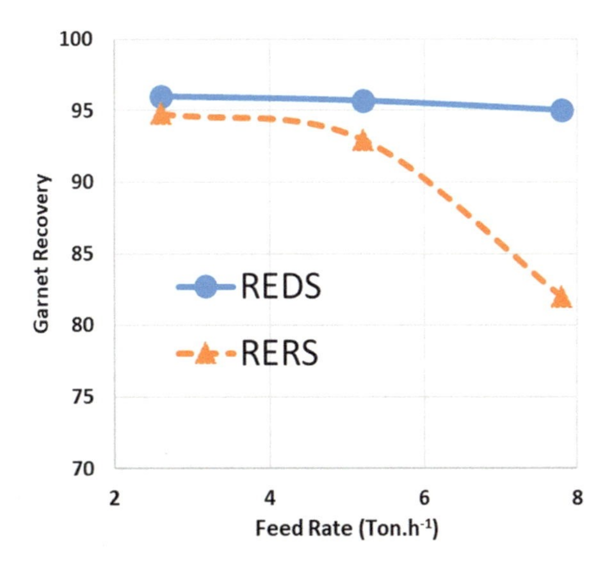

Fig. 2.11 Performance evaluation of the RED and RER separators [30]

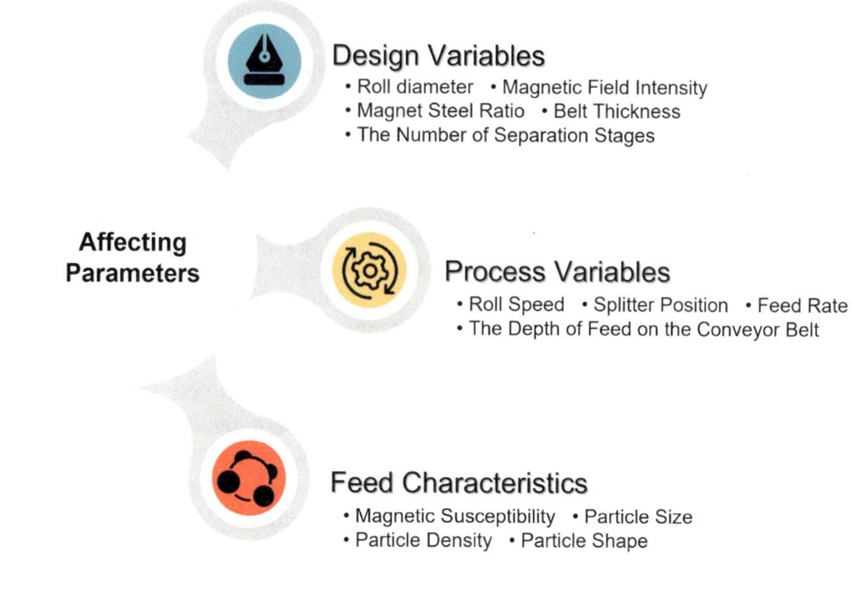

Fig. 2.12 Parameters affecting the performance of RERS [13]

Even though RERS is usually utilized in dry mode, its wet usages have also been reported, for instance, in the concentration of diamonds [33]. RERS are extensively utilized in various industries and applications for both pre-concentration and ultimate cleaning purposes. The highly effective separation of high-intensity RERS has made them preferred in many applications to other magnetic machines. It has been widely utilized as an effective separator for iron oxide removal from various materials such as ceramic, glass, abrasives, calcined magnesite, and chemicals. Additionally, it is reported that RERS can be a suitable separator for recovering zircon and rutile from non-magnetic minerals in the sand deposits. Coal Separation and waste processing can also be mentioned as other applications of RERS [13].

2.3.3 Induced Magnetic Roll Separator

The induced magnetic roll separator (IMRS) comprises a rotating laminated roll manufactured of several magnetizable and non-magnetic discs alternately placed next to each other (Fig. 2.13). The induction system can provide a magnetic field of approximately 2 T [34].

The magnetic roll is installed between poles of an electromagnet which are shaped pieces. The electromagnet causes inducing a magnetic field that generates areas of the strong magnetic field gradient. A vibratory feeder moves a controlled thin stream of materials to the top of the roll. As the induced roll rotates, the particles pass through

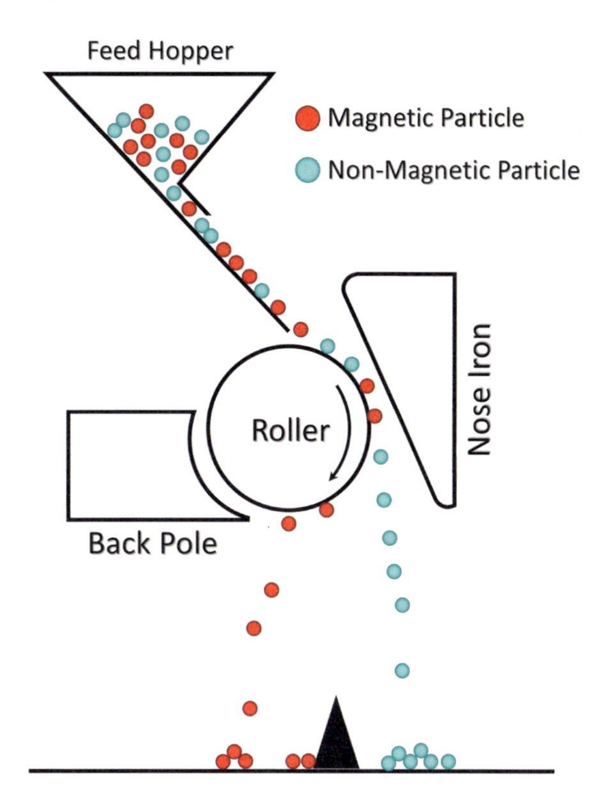

Fig. 2.13 Schematic diagram of a IMRS [34]

a narrow passage between the pole and the magnet surface [22, 35]. Non-magnetic materials are not affected by the magnetic field. Therefore, they are discharged from the roll, whereas the magnetic ones are attracted to the magnetic roll. Subsequently, when magnetic particles come out of the magnetic field, they are collected in a certain bin. It should be accentuated that prior to feeding to an IMRS, ferromagnetic materials have to be eliminated with another appropriate magnetic separator; otherwise, they will accumulate in the gap between the roll and pole, causing the separation process to stop [36]. The gap and splitter location are both adjustable, which is very important in operating the separation process. The generated magnetic induction in the gap depends on the magnetizing coils current and gap width. For having an effective separation in IMRS, some points must be considered carefully:

- The input materials have to be dry
- A free-flowing material stream is required
- The particle size should be from 100 μm to 2 mm
- The gap should be set to at least 2.5 times the average grain size
- The throughput can range from 1 to 3 t/h per meter of the roll [13, 36, 37].

IMRSs have been widely employed for beach sands beneficiation and the elimination of feebly magnetic impurities from feldspar, glass, magnesite, wollastonite, and other industrial minerals. This machinery can also be applied for the concentration of minerals such as monazite, chromite, and wolframite [22, 38]. Although numerous applications, much simpler to design and easier to use, permanent magnetic rolls have replaced IMRSs in some applications. This is mainly because there are a number of limitations in the separation process using this IMRS, such as limited particle size range and large mass, and comparatively low capacity. However, efforts to develop and improve the performance of these devices have not stopped [13]. As an example, a laboratory-scale induced roll magnet was developed by Outokumpu Technology to separate particles up to 1 mm. The main characteristic of the separator was the magnetically lifting (against gravity) mechanism, resulting in highly selective separation. Since undesirable particles are not entrapped in this mechanism, this separator is an appropriate option, especially when the purpose of separation is to increase the grade of non-magnetic concentrates [1].

2.3.4 *Lift Roll Magnetic Separator (LRMS)*

Lift Roll Magnetic Separator can magnetically separate materials. During separation, the feed particles are exposed to a magnetic field applied by a magnetically induced roll. While magnetic particles are influenced and attracted by magnetic force, non-magnetic would not attract and accumulate in the non-magnetic portion (Fig. 2.14). Similar to the rolling magnet developed by Outokumpu, magnetic materials are lifted and separated from the material stream against the gravitational force. This causes minimizing the entrapped non-particle materials, leading to producing a fresh magnetic product [13, 39].

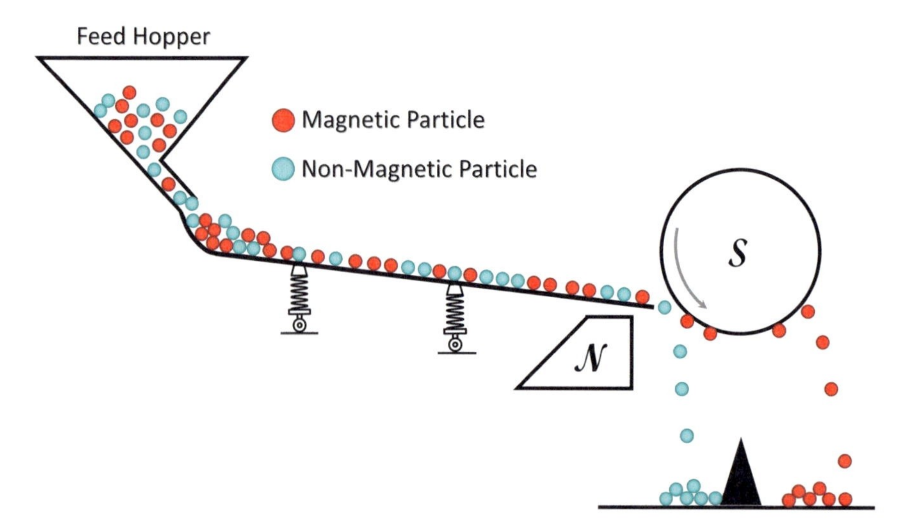

Fig. 2.14 Separation process in a lift roll magnetic separator [39]

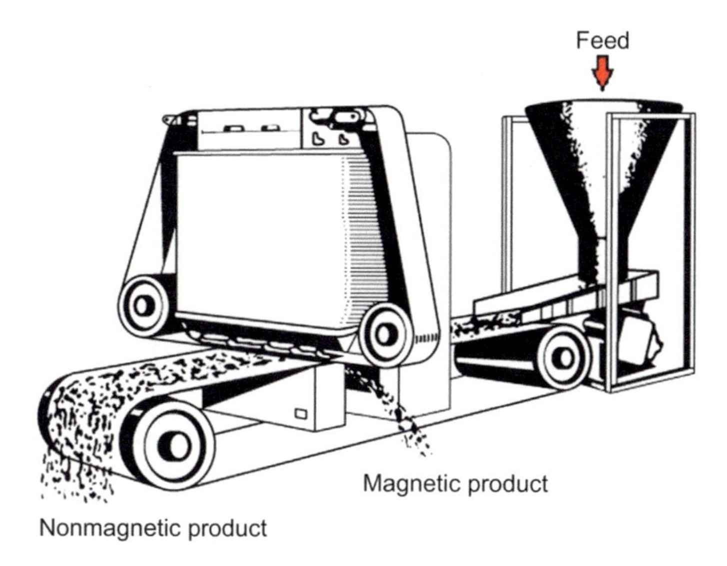

Fig. 2.15 Cross-belt magnetic separator [1]

The selectivity of separation by LRMS is higher than IMRS. However, the greatest weakness of this separator is its limited capacity. In the case of separation rutile from leucoxene and garnet from monazite, LRMS can be effectively utilized, and the optimum feed rate could be from 1.5 to 2 t/h [13].

2.3.5 Cross-Belt Magnetic Separator

Cross-belt (Fig. 2.15) and disc-belt magnetic separators, one of the oldest types of separators, once have been extensively employed in the mineral sands industry and weakly magnetic ores. However, rare-earth magnet drums and rolls have replaced them. In this high-intensity electromagnetic separator, the material stream is fed to the machine in a single layer on the conveyer belt. The loaded conveyor belt passes between electromagnet poles. The lower pole is fixed and flat, whereas the upper one is a shaped pole that is movable and can be raised or lowered when required. The upper pole position is regulated based on input material size. It should be positioned to be at least 2.5 times the size of the largest particle present in the material stream. Cross-belt magnetic separators can be used for separating materials ranging from 75 μm to 4 mm. Also, the width of the conveyor belt can vary from 100 to 600 mm. The operating magnetic induction changes depend on the distance between the poles and can reach 2 T. Though using a cross-belt separator, several types of magnetic products can be recovered in one pass, and the low capacity is its main drawback [1, 40].

2.4 High-Gradient Magnetic Separator (HGMS)

As its name implies, a high gradient magnetic separator employs magnetic field charges to separate magnetic materials. Marston and Kolm introduced this method at the Massachusetts Institute of Technology in the late 1960s and early 1970s. In this electromagnet-based technology, a uniform high-intensity magnetic field is generated, and sharp-edged ferromagnetic materials are located in the magnetic field, causing to generating high field gradients [41, 42]. HGMS makes it possible to efficiently separate micron-sized, very weakly magnetic materials. This is why this technology has attracted considerable theoretical and experimental attention in various industries such as mineral processing, coal beneficiation, nanotechnology, and medicine [41, 43, 44]. In mineral processing, HGMS has traditionally been employed in kaolin clay benefaction. It is also used for iron removal from process streams in steel and power plants as well as wastewater treatment and solids through magnetic seeding [45, 46].

In mining and mineral processing, coal beneficiation is the industry in which most research on HGMS efficiency has been conducted. The first satisfactory results were reported in 1977, and subsequently, many investigations have been carried out to investigate the efficiency of HGMS in dry coal beneficiation. Obtained results for the use of the HGMS technique in coal beneficiation considerably vary among the different investigations. However, what has been reported by all these investigations is that the performance of HGMS is deeply dependent on the range of particle size being treated. It is concluded that the most efficient separation occurs when the amount of fine particles (-150 μm) in feed is minimized [3]. Even though the investigations have concluded that there is potential for using HGMS in coal beneficiation, further research is needed to better understand the complexities of this type of separator and improve its performance on an industrial scale [4].

2.5 Main Applications and Producers

In general, all these methods and systems have been used in different mineral processing investigations, which are summarized in Tables 2.3 and 2.4. The representative producers of magnetic separators are presented in Table 2.5.

Table 2.3 Summary of the IMRS dry magnetic separators' application for various mineral beneficiation

Feed	Mine location	Variable studied	Application/purpose	Remarks	References
Nepheline-Syenite Ores	Turkey	– Rotor speed – Magnetic field – Feed rate – Feed particle size	Rejecting iron-bearing impurities	– It was possible to produce clean concentrates containing 0.24 and 0.28% iron oxide from original samples containing 6.0% and 5.3% Fe_2O_3, respectively	[47]
Ferruginous Low-Grade Manganese Ore	India	– Magnetic intensity – Rotor speed	Used as two stages separation in the flowsheet	– Classification followed by two-stage high-intensity magnetic separation (1.7 and 1.1 T) process can recover 35–40% material of ferromanganese grade with 47–49% Mn recovery	[48]
Ferruginous Manganese Ore	India	– Magnetic intensity – Particle size – Rotor speed – Splitter position	Rejecting iron-bearing gangue minerals	– The proposed models can be used to modify the operational parameters at mineral beneficiation plants to minimize the effects of variations in the raw material characteristics	[49]
Chromite Beneficiation Plant Tailing	Turkey	– Magnetic intensity – Rotor speed – Feed rate	Used as a cleaner after a gravity separator	– An IMRS was used to produce a saleable chromite concentrate from the pre-concentrated output of a gravity separator. As the outcome of two-stage consecutive concentrations, a saleable chromite concentrate including 42.9% Cr_2O_3 with 73.5% recovery could be produced	[50]

(continued)

Table 2.3 (continued)

Feed	Mine location	Variable studied	Application/purpose	Remarks	References
Ferruginous-Chromite Ore	India	– Separation Method – Magnetic intensity – Rotor speed	Used as a cleaner to recover the values	– The product can be enriched to having Cr:Fe ratio of 2.31 and 3.35 by a gravity concentration (wet shaking table) and an IMRS respectively. Consequently, IMRS was suggested for the quality improvement of the examined off-grade chromite ore	[51]
Hematite	India	– Rotor speed – Magnetic field – Feed rate	Separating hematite fines from the low-grade fines	– Based on the proposed model, it can be concluded that, IMRS is one of the suitable equipment to separate or pre-concentrate para-magnetic minerals like hematite. Further, it is possible to improve the grade by improving the liberation of hematite by reducing the particle size to <1 mm	[22]
Garnet Fines	India	– Magnetic intensity – Rotor speed – Feed rate – Splitter position – Particle size	Investigation of the role of various parameters in separation process	– The results show that segregation of the coarser particle is segregated in non-magnetic fraction, while finer sized particles are at the magnetic fraction	[52]

(continued)

Table 2.3 (continued)

Feed	Mine location	Variable studied	Application/purpose	Remarks	References
Low-Grade Iron Ore	Sudan	– Magnetic intensity – Rotor speed – Particle size	Used as rougher and cleaner in the flowsheet	– Using a two-stage separation, roughing and cleaning, it was possible to obtain a high-grade concentrate assaying about 64% Fe at a recovery of 72% from a feed with 36% Fe and 48% silica	[53]
Chromite Beneficiation Plant Tailing	India	– Magnetic intensity – Rotor speed	improve the Cr:Fe ratio	– Based on the obtained results, two flowsheets comprising gravity, magnetic separation and flotation, were used to recover chromite beneficiation plant tailing. A chromite concentrate of 45.0% Cr_2O_3 with a Cr:Fe ratio of 2.3 can be produced from the tailing analyzing 17.0% Cr_2O_3 and Cr:Fe ratio of 0.49	[38]
Hematite	India	– Magnetic intensity – Rotor speed – Feed rate – Splitter position – Particle size	Separate hematite fines from the low-grade fines	– It was concluded that a IMRS along with a RERMS can be utilized as a pre-concentrator/scavenger unit to discard a maximum amount of gangue in a dry operation which can save energy regarding the fine grinding of the ore for liberation	[37]

(continued)

Table 2.3 (continued)

Feed	Mine location	Variable studied	Application/purpose	Remarks	References
Magnesite-Dolomite Ore	Egypt	– Magnetic intensity – Rotor speed	Separate iron contained magnesite from dolomite	– Using an IMRS, a magnesite concentrate product with 46.02% recovery is attainable. This magnesite product is suitable for basic refractory. Besides, middling fraction was produced, containing 44.63% magnesite and 10.95% dolomite This product is suitable for making low-loss forsterite dielectrics	[54]
Manganese Ore	India	– Particle size – Magnetic intensity – Roll speed	Improving Mn:Fe ratio by desliming and subsequent concentration	– Based on the obtained results, a flowsheet for the beneficiation of low-grade manganese ore was proposed. It comprises a density separator as a pre-concentrator followed by two-stage magnetic separation. The overall recovery of manganese in the final product from the is 44.7% with an assay value of 45.8% and the Mn:Fe ratio of 3.1	[55]

Table 2.4 Summary of the RERS dry magnetic separators' application for various mineral beneficiation

Feed	Mine location	Variable studied	Application/purpose	Remarks	Ref
Ilmenite	India	– Feed size	Separating the ilmenite along with other magnetics	– On processing this feed to recover ilmenite by using spirals followed by RERS, the recovery of 94% was obtained	[56]
Bauxite ore	India	– Roll speed – Feed size	Separating paramagnetic minerals, namely hematite, goethite, and rutile	– The best separation result was achieved at the roll speed of 3 rpm on the 350–690 μm size fraction. Fe_20_3 and TiO_2 contents could be reduced to 1.52 and 5.16% in the concentrate from their respective values of 3.31 and 7.31% in the feed	[57]
Coal	UK	– Feed size	Separating pyrite from coal	– RERS removed both ash and sulfur from the coal. Ash removal was maximized at 40% for the 106–500 μm coal size fraction, with total sulfur removal being in the order of 10%	[58]
Lignite Fly Ash	Turkey	– Roll speed – Feed size	Used to discharging iron-bearing impurities	– A flowsheet is developed on the basis of differences in gravities and magnetic susceptibilities of the byproducts	[59]
Manganese Ore	Turkey	– Feed size	Used for separation of braunite	– Results showed that using RERS, metamorphic braunite-rich manganese deposits can be successfully enriched at high Mn recovery with the best results in 1–10 mm size range	[60]

(continued)

Table 2.4 (continued)

Feed	Mine location	Variable studied	Application/purpose	Remarks	Ref
Silica Sand	Egypt	– Feed size – Belt speed – Splitter angle	Removal of Fe_2O_3	– It was concluded that the method of feeding to the RER magnetic separator affects the grade of the final sand concentrate – In case of separating the whole bulk, without fractionation, the sand concentrate assays 0.083% Fe_2O_3 (56.56% removal), but with fractionation, this reaches 0.026% Fe_2O_3 with (85.34% removal) – The quality of the silica sand concentrate produced satisfies the requirement for both colorless glass and insulating glass fibers production	[30]
Trona	USA	– Feed size – Splitter position	Used to pre-concentrate the trona for the removal of dolomitic shale and illite	– The purity of trona concentrate is limited by the particle size and impurity content of the trona ore – A trona concentrate of 91.7% purity with a recovery 91.1% can be obtained by two-stage magnetic separation, which could be further improved to 98.1% purity at a recovery of 83.8% by subsequent flotation of the non-magnetic trona product	[61]

(continued)

Table 2.4 (continued)

Feed	Mine location	Variable studied	Application/purpose	Remarks	Ref
Nepheline Syenite	Egypt	– Belt speed – Feed size – Feed rate	Removing the iron impurities	– It was possible to produce clean concentrates containing 0.24% and 0.28% iron oxide from original samples containing 6.0% and 5.3% Fe_2O_3, respectively. These products are acceptable as glass grade materials in amber glass and fiberglass industries, as well as for ceramic purposes	[62]
Bauxite ore	Greece	– Feed size	Used to separate the calcite from bauxite	– RERS can be used to remove unwanted lime stone from bauxite. This process can upgrade existing stockpiles of −20 mm or −30 mm fractions and also encourage mining of lower grade bauxite deposits – The advantage of using RERS compared with heavy media separation, which is using in Greek mines, is the low capital investment and operating cost	[63]
Ferruginous Chromite Ore	India	– Roll speed – Feed rate – Separator type	Separating ferruginous gangue minerals from two different ores	– Among the dry magnetic separators, RERS can be used for both the ores to improve the Cr:Fe ratio economically	[12]

Table 2.5 The main magnetic separator producers

Equipment	Manufacturer	Location
Magnetic separator	Andritz	Austria
	Bakker magnetics	Netherlands
	Buhler	Switzerland
	Bunting	UK
	Dragoelectronica	Spain
	Eaton	USA
	Eldan recycling	Denmark
	Erga	Russia
	Eriz	UK
	Filtra	USA
	Hammel	Germany
	Huate	China
	JSYY	China
	Longi	China
	Losma	Italy
	Magengine	China
	Magnetense	Italy
	MDR	Slovenia
	Metso	Finland
	Otech	Germany
	Outotec	Finland
	ProTechnica	Bulgaria
	Regulator Cetrisa	Spain
	SDM	China
	Sesotech	Germany
	Sinosteel	China
	Trotech	Austria

References

1. Wills' Mineral Processing Technology. Elsevier (2016)
2. Karmazin, V.V.: Theoretical assessment of technological potential of magnetic and electrical separation. Magn. Electr. Sep. **8**, 139–159 (1997). https://doi.org/10.1155/1997/23759
3. Dwari, R.K., Rao, D.S., Reddy, P.S.R.: Magnetic separation studies for a low grade siliceous iron ore sample. Int. J. Min. Sci. Technol. **23**, 1–5 (2013). https://doi.org/10.1016/j.ijmst.2013.01.001
4. Dwari, R.K., Rao, K.H.: Dry beneficiation of coal—a review. Miner. Process. Extr. Metall. Rev. **28**, 177–234 (2007). https://doi.org/10.1080/08827500601141271
5. Dean, Reginald Scott, and C.W.D.: Magnetic separation of ores. US Government Printing Office (1941)

6. Oberteuffer, J.A.: Magnetic separation: a review of principles, devices, and applications. IEEE Trans. Magn. **10**, 223–238 (1974). https://doi.org/10.1109/TMAG.1974.1058315
7. Bhoja, S.K., Tripathy, S.K., Murthy, Y.R., Ghosh, T.K., Raghu Kumar, C., Chakraborty, D.P.: Influence of mineralogy on the dry magnetic separation of ferruginous manganese ore—a comparative study. Minerals **11**, 1–20 (2021). https://doi.org/10.3390/min11020150
8. Venkatraman, P., Knoll, F.S., Lawver, J.E.: Magnetic and electrostatic separation. Miner. Process. Des. Oper. 629–687 (2016). https://doi.org/10.1016/b978-0-444-63589-1.00017-4
9. Chehreh Chelgani, S., Leißner, T., Rudolph, M., Peuker, U.A.: Study of the relationship between zinnwaldite chemical composition and magnetic susceptibility. Miner. Eng. **72**, 27–30 (2015). https://doi.org/10.1016/J.MINENG.2014.12.024
10. Dobbins, M., Domenico, J., Dunn, P.: A discussion of magnetic separation techniques for concentrating ilmenite and chromite ores (2007)
11. Ezhov, A.M., Shvaljov, Y.B.: Dry magnetic separation of iron ore of the bakchar deposit. Procedia Chem. **15**, 160–166 (2015). https://doi.org/10.1016/j.proche.2015.10.026
12. Tripathy, S.K., Murthy, Y.R., Singh, V., Suresh, N.: Processing of ferruginous chromite ore by dry high-intensity magnetic separation. Miner. Process. Extr. Metall. Rev. **37**, 196–210 (2016). https://doi.org/10.1080/08827508.2016.1168418
13. Tripathy, S.K., Banerjee, P.K., Suresh, N., Murthy, Y.R., Singh, V.: Dry high-intensity magnetic separation in mineral industry—a review of present status and future prospects. Miner. Process. Extr. Metall. Rev. **38**, 339–365 (2017). https://doi.org/10.1080/08827508.2017.1323743
14. Wang, F., Tang, D., Gao, L., Dai, H., Wang, P., Gong, Z.: Magnetic entrainment mechanism of multi-type intergrowth particles for low-intensity magnetic separation based on a multiphysics model. Miner. Eng. **149**, 106264 (2020). https://doi.org/10.1016/j.mineng.2020.106264
15. Desportes, H.: Three decades of superconducting magnet development. Cryogenics (Guildf). **34**, 47–56 (1994). https://doi.org/10.1016/S0011-2275(05)80009-8
16. Morgan, D.G., Bronkala, W.J.: The selection and application of magnetic separation equipment. Part II. Magn. Electr. Sep. **4**, 151–172 (1993). https://doi.org/10.1155/1993/81597
17. Morgan, D.G., Bronkala, W.J.: The selection and application of magnetic separation euipment. Part I. Magn. Electr. Sep. **3**, 5–16 (1991). https://doi.org/10.1155/1991/26791
18. Verma, A., Pandey, O.P., Sharma, P.: Ferrite permanent magnet—an overview. Indian J. Eng. Mater. Sci. **7**, 364–369 (2000). https://doi.org/10.1002/chin.200138258
19. SOLLAU CZ, s. r. o. Available online: https://www.sollau-cz.com/. Accessed on 30 Aug 2021
20. Qawaqzeh, M.Z.: Design and computation of a suspended magnetic separator for processing metallurgic slag. Prz. Elektrotechniczny **96**, 67–72 (2020). https://doi.org/10.15199/48.2020.04.13
21. Meireles, M., Bourgeois, F., Tourbin, M., Guiraud, P., Frances, C., Meireles, M., Bourgeois, F., Tourbin, M., Guiraud, P., Review, C.F.: Removal of oversize & recovery of particles from suspensions in the nano size range (2015)
22. Tripathy, S.K., Banerjee, P.K., Suresh, N.: Separation analysis of dry high intensity induced roll magnetic separator for concentration of hematite fines. Powder Technol. **264**, 527–535 (2014). https://doi.org/10.1016/J.POWTEC.2014.05.065
23. Arvidson, B.R., Henderson, D.: Rare-earth magnetic separation equipment and applications developments. Miner. Eng. **10**, 127–137 (1997). https://doi.org/10.1016/S0892-6875(96)00139-2
24. Marinescu, M., Marinesco, N., Unkelbach, K.H., Schnabel, H.G., Hock, S., Krammig, H., Wagner, R., Zoller, R.: New permanent magnetic separator with NdFeB meets theoretical predictions. IEEE Trans. Magn. **25**, 2732–2738 (1989). https://doi.org/10.1109/20.24516
25. Wasmuth, H.D.: New medium-intensity drum-type permanent magnetic separator PERMOS and its practical applications for the processing of industrial minerals and martitic iron ores. Magn. Electr. Sep. **6**, 201–212 (1995). https://doi.org/10.1155/1995/94934
26. Wells, I.S., Rowson, N.A.: Application of rare earth magnets in mineral processing. Magn. Electr. Sep. **3**, 105–111 (1992). https://doi.org/10.1155/1992/16139
27. Arvidson, B.R.: Recent developments of rare earth magnetic roll and drum separators. Proc. Soc. Miner. Eng. Annu. Meet. Salt Lake City, UT, USA (2000)

28. Anderson, C.G., Dunne, R.C., Uhrie, J.L.: Mineral processing and extractive metallurgy: 100 years of innovation. 698
29. Norrgran, D.A., Marin, J.A.: Rare earth permanent magnet separators and their applications in mineral processing. Min. Metall. Explor. **11**(1), 41–45 (1994). https://doi.org/10.1007/BF03403039
30. Ibrahim, S.S., Farahat, M.M., Boulos, T.R.: Optimizing the performance of the RER magnetic separator for upgrading silica sands. Part. Sci. Technol. **35**, 21–28 (2017). https://doi.org/10.1080/02726351.2015.1121179
31. Arvidson, B.R.: Advances in rare earth magnetic drum separators for heavy minerals sands processing. SAIMM Heavy Minerals Conference, Durban, South Africa (1999)
32. Orhan, E., Gülsoy, O.: Importance of magnet-steel configuration in dry high intensity permanent magnetic rolls : theoretical and practical approach. Physicochem. Probl. Miner. Process. **38**, 301–309 (2004)
33. Rylatt, M.G.: Diamond processing at Ekati in Canada. Min. Eng. (1999)
34. Dolgopolov, O.A.: A new generation of magnetic roll separators based on rare-earth magnets for concentration and purification of weakly magnetic materials. Glas. Ceram. **62**(5), 155–156 (2005). https://doi.org/10.1007/S10717-005-0061-X
35. Singh, V., Nag, S., Tripathy, S.K.: Particle flow modeling of dry induced roll magnetic separator. Powder Technol. **244**, 85–92 (2013). https://doi.org/10.1016/J.POWTEC.2013.03.053
36. Premaratne, W.A.P.J., Rowson, N.A.: The processing of beach sand from Sri Lanka for the recovery of titanium using magnetic separation. Phys. Sep. Sci. Eng. **12**, 13–22 (2003). https://doi.org/10.1080/1478647031000101232
37. Tripathy, S.K., Singh, V., Rama Murthy, Y., Banerjee, P.K., Suresh, N.: Influence of process parameters of dry high intensity magnetic separators on separation of hematite. Int. J. Miner. Process. **160**, 16–31 (2017). https://doi.org/10.1016/J.MINPRO.2017.01.007
38. Tripathy, S.K., Murthy, Y.R., Singh, V.: Characterisation and separation studies of Indian chromite beneficiation plant tailing. Int. J. Miner. Process. **122**, 47–53 (2013). https://doi.org/10.1016/J.MINPRO.2013.04.008
39. Zong, Q.X., Fu, L.Z., Bo, L.: Variables and applications on dry magnetic separator. E3S Web Conf. **53**, 02019 (2018). https://doi.org/10.1051/E3SCONF/20185302019
40. Arvidson, B.R.: The many uses of rare-earth magnetic separators for heavy mineral sands processing. Int. Heavy Miner. Conf. (2001)
41. Moeser, G.D., Roach, K.A., Green, W.H., Hatton, T.A., Laibinis, P.E.: High-gradient magnetic separation of coated magnetic nanoparticles. AIChE J. **50**, 2835–2848 (2004). https://doi.org/10.1002/aic.10270
42. Hise, E.C., Holman, A.S.: Development of high-gradient and open-gradient magnet separation of dry fine coal. IEEE Trans. Magn. **17**, 3314–3316 (1981). https://doi.org/10.1109/TMAG.1981.1061625
43. Ritter, J.A., Ebner, A.D., Daniel, K.D., Stewart, K.L.: Application of high gradient magnetic separation principles to magnetic drug targeting. J. Magn. Magn. Mater. **280**, 184–201 (2004). https://doi.org/10.1016/j.jmmm.2004.03.012
44. Oberteuffer, J.A.: High gradient magnetic separation. IEEE Trans. Magn. **9**, 303–306 (1973)
45. Ying, T.Y., Yiacoumi, S., Tsouris, C.: High-gradient magnetically seeded filtration. Chem. Eng. Sci. **55**, 1101–1113 (2000). https://doi.org/10.1016/S0009-2509(99)00383-8
46. Ge, W., Encinas, A., Araujo, E., Song, S.: Magnetic matrices used in high gradient magnetic separation (HGMS): a review. Results Phys. **7**, 4278–4286 (2017). https://doi.org/10.1016/j.rinp.2017.10.055
47. Ibrahim, S.S., Mohamed, H.A., Yedilbayev, B.A.: Dry magnetic separation of magnetite ores. Physicochem. Probl. Miner. Process. **36**, 173–183 (2002). https://doi.org/10.52571/ptq.v17.n34.2020.724_p34_pgs_700_710.pdf
48. Singh, V., Ghosh, T.K., Ramamurthy, Y., Tathavadkar, V.: Beneficiation and agglomeration process to utilize low-grade ferruginous manganese ore fines. Int. J. Miner. Process.**99**, 84–86 (2011)

49. Singh, V.: Connectionist approach for modeling the dry roll magnetic separator. Min. Metall. Explor. **26**(3), 127–132 (2009). https://doi.org/10.1007/BF03402225
50. Aslan, N., Kaya, H.: Beneficiation of chromite concentration waste by multi-gravity separator and high-intensity induced-roll magnetic separator. Arab. J. Sci. Eng. **34**, 285–297 (2009)
51. Tripathy, S.K., Singh, V., Ramamurthy, Y.: Improvement in Cr:Fe Ratio of Indian Chromite Ore for Ferro Chrome Production. Undefined (2012)
52. Tripathy, S.K., Suresh, N.: Influence of particle size on dry high-intensity magnetic separation of paramagnetic mineral. Adv. Powder Technol. **28**, 1092–1102 (2017). https://doi.org/10.1016/j.apt.2017.01.018
53. Seifelnassr, A.A.S., Moslim, E.M., Abouzeid, A.-Z.M.: Concentration of a Sudanese low-grade iron ore. Int. J. Miner. Process. **122**, 59–62 (2013). https://doi.org/10.1016/J.MINPRO.2013.04.001
54. Yehia, A., Al-Wakeel, M.: Role of ore mineralogy in selecting beneficiation route for magnesite-dolomite separation. Physicochem. Probl. Miner. Process. **49**, 525–534 (2013). https://doi.org/10.5277/PPMP130213
55. Tripathy, S.K., Banerjee, P.K., Suresh, N.: Effect of desliming on the magnetic separation of low-grade ferruginous manganese ore. Int. J. Miner. Metall. Mater. **22**(7), 661–673 (2015). https://doi.org/10.1007/S12613-015-1120-0
56. Babu, N., Vasumathi, N., Rao, R.B., Babu, N., Vasumathi, N., Rao, R.B.: Recovery of ilmenite and other heavy minerals from teri sands (red sands) of Tamil Nadu, India. J. Miner. Mater. Charact. Eng. **8**, 149–159 (2009). https://doi.org/10.4236/JMMCE.2009.82013
57. Bhagat, R., Banerjee, B., Saha, P., Mukherjee, B.: Dry magnetic separation of bauxite ore. Undefined (2006)
58. Saeid, A.M., Butcher, D.A., Rowson, N.A.: Coal desulphurisation and ash removal in intensified magnetic fields. Magn. Electr. Sep. **4**, 107–116 (1993). https://doi.org/10.1155/1993/10494
59. Özdemir, O., Çelik, M.S.: Characterization and recovery of lignitic fly ash byproducts from the tuncbilek power station. Can. Metall. Q. **41**, 143–150 (2013). https://doi.org/10.1179/CMQ.2002.41.2.143
60. Grieco, G., Kastrati, S., Pedrotti, M.: Magnetic enrichment of braunite-rich manganese ore at different grain sizes. Min. Process. Extr. Metall. Rev. **35**, 257–265 (2013). https://doi.org/10.1080/08827508.2013.793680
61. Ozdemir, O., Gupta, V., Miller, J.D., Çınar, M., Çelik, M.S.: Production of trona concentrates using high-intensity dry magnetic separation followed by flotation. Min. Metall. Explor. **28**(2), 55–61 (2011). https://doi.org/10.1007/BF03402388
62. Ibrahim, S.S., Mohamed, H.A., Boulos, T.R.: Dry magnetic separation of nepheline syenite ores. Physicochem. Probl. Miner. Process. **36**, 173–183 (2002)
63. Stamboliadis, E.T., Kailis, G.: Removal of limestone from bauxite by magnetic separation. Eur. J. Miner. Process. Environ. Prot. **4**, 84–90 (2004)

Chapter 3
Gravity Separation

3.1 Introduction

Gravity separation is a method in which materials/minerals are separated based upon the difference in particles density [1]. Water as a medium is an important factor in improving the differential movement between various particles during gravity separation and eventually sorting different particles [2]. When water is removed, and air plays the role of separation medium, the separation method is defined as dry, pneumatic, or air gravity separation [3]. In general, dry gravity separation has the attraction of decreasing capital and operating costs compared to the wet mode. Additionally, no need for water, chemicals, and drying procedures has made it an environmentally friendly process [4].

A broad range of devices can be used based on this aspect, such as jigs, tables, spirals, and heavy-media separation devices [5]. Density-based concentration techniques were initially welcomed due to their low negative environmental impact and relative simplicity compared to other separation methods. However, in the early twentieth century, the importance of this method declined because of the flotation separation development. This occurred because not only can flotation cover the effective range of gravity concentration methods (Fig. 3.1), but it can also be utilized for separating finer particles [6]. Of course, it is proven that some modern gravity separators, when are used in conjunction with the enhanced pumping systems and instrumentation, can be efficient for the concentration of fine particles (down to 50 μm) [7]. Recently, rising flotation costs have led some mineral processing plants to reevaluate density-based separation techniques. Nowadays, this method is the major concentrating procedure for tin, tungsten, and iron and is extensively utilized for treating gold, coal, and beach sands [6].

In density-based concentration techniques, the separation process occurs by the relative movement of mineral particles due to different forces, including gravity and one or more other forces. The latter is usually resistant to the movement of particles and is provided by a fluid medium, namely air or water [6]. For an effective separation, there must be a significant density differential between the gangue and

S. C. Chelgani and A. Asimi Neisiani, *Dry Mineral Processing*,
https://doi.org/10.1007/978-3-030-93750-8_3

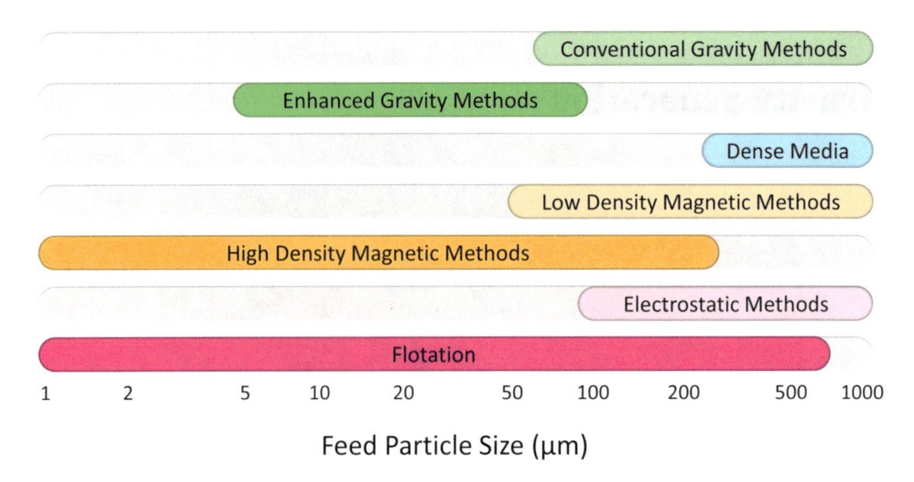

Fig. 3.1 The effective size range of separation methods [6]

valuable minerals. In density-based separation, concentration criterion ($\Delta\rho$) can be defined as a criterion for determining the feasibility and ease of separation (Eq. 3.1):

$$\Delta\rho = \frac{\rho_h - \rho_f}{\rho_l - \rho_f} \tag{3.1}$$

where ρ_h, ρ_l, and ρ_f are the specific density of the heavy particle, light particle, and fluid medium, respectively [8]. Generally, the gravity concentration process is comparatively easy when the magnitude of the concentration criterion is greater than 2.5. The lower the value of this factor, the lower the separation efficiency. Moreover, by increasing this ratio, the separation process will be sharper [9]. For instance, in the case of separating gold from quartz using water as a fluid medium, the specific density of quartz and gold are 2.650 and 19.300 kg m^{-3}, respectively. Based on these values, the concentration criterion is calculated to be 11.1, and therefore, this is not unexpected that the gravity separation of gold can be performed quite effectively [6]. If this ratio is calculated to separate gold particles from quartz in the air medium, the value will be 7.3 [8].

The movement of mineral particles in a fluid medium, whether in air or water, is dependent not only on their specific density but also on their size. Actually, a particle has to be coarse enough for Newton's force to move it (Eq. 3.2). The larger the particle size, the more it is affected in the separation process; thus, the separation efficiency will be enhanced [6].

$$v = \left[\frac{3gd(\rho_p - \rho_f)}{\rho_f}\right]^{\frac{1}{2}} \tag{3.2}$$

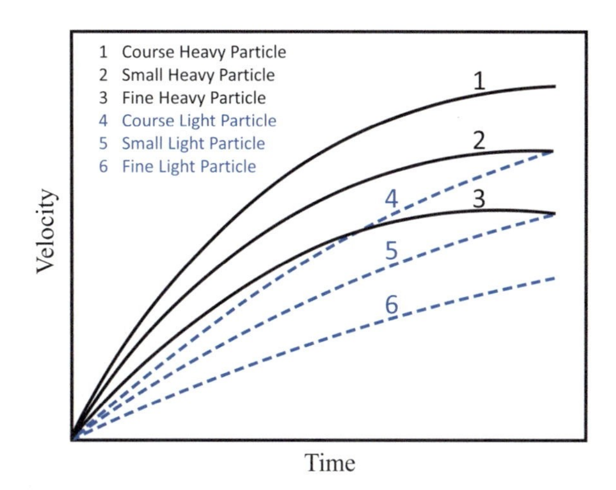

Fig. 3.2 Settling velocities of different particles [8]

where:

- v velocity of the particle
- g gravitational constant
- ρ_p particle density
- d particle diameter.

To better understand the simultaneous effect of particle density and particle size on gravity separation, the variation of settling velocities of different particles over time is shown in Fig. 3.2 [8]. Consequently, the very fine mineral particle that their movement is chiefly affected by surface friction cannot be effectively concentrated by conventional gravity methods. For overcoming this drawback, enhanced gravity concentrators have been developed. Centrifugal force can decrease the dependence of particle motion on the size and make their relative motion gravity-dependent. [6].

Water-based gravity separation techniques are much more typically considered compared to dry ones. However, the requirement for a significant amount of water and water scarcity in many regions of the world have led to dry methods being considered these days [10]. For example, throughout wet jigging of each ton of coal, approximately 5 tons of process water is needed, and fresh water has to be incessantly added to the system [11]. Additionally, dry gravity concentration methods are preferable for some processes, such as coal beneficiation. The difficulty of slime management in wet processes can be the main reason, specifically for those plants are located in cold or arid regions. However, the main advantage of wet gravity concentration methods is their high efficiency over dry methods, making them still favorable in most industries [12].

More than any other process in mineral processing, coal beneficiation has benefited from dry gravity separation [13]. Air-based separation methods of coal cleaning can be generally categorized into two main groups, namely air dense medium and settling devices. Air dense medium fluidized bed separator (ADMFBS) and Counter-current fluidized cascade (CCFC) can be mentioned as air dense medium devices.

The settling process can also be further divided into air table separation and dry jigging [13]. In the following dry density-based separation methods are investigated.

3.2 Air Jig

As one of the oldest preconcentration and concentration processes, jigging has been utilized for many years. This density-based separation method is still extensively used in mineral processing plants. Until recently, it has been used in conjunction with dense medium separation (DMS) for coarse-sized ores treatment. Currently, jigging devices have exceeded the limits of mineral processing procedures, being applied in other processes like recycling plants. Since water scarcity has raised global concerns, waterless jigging has been noticed these days [14].

Dry jigging, also known as air jigging or pneumatic jigging, is a method in which pulsating air causes heavy mineral particles to separate from light ones. The efficiency of this method, which has been applied and industrialized for coal beneficiation in many countries, principally depends on the differences in the characteristics of gangue and valuable particles such as density, particle size, and shape factor [15]. Several investigations have indicated that this technique is successful and can be utilized not only for coal beneficiation [16] but also for sand and ferrous minerals [17, 18]. This concept was initially introduced in the coal beneficiation application in the 1920s when it was involved in wet jigging to develop pneumatic cleaning devices. After that, considerable enhancements have been achieved in its design and operation [19].

Dry jigging devices (Fig. 3.3) comprise an inclined perforated plate on which the feed is moved, using a vibration driver, and exposed to two independent upward air streams [14]. The function of the first stream, which is a continuous airflow, is to loosen the material bed and results in a uniformly distributed airflow. Additionally, a pulsed air stream improves the density stratification of particles in the bed when moving on the dry jig deck. The cumulative impact of the two distinguished air streams permits accurate control of frequency, and therefore, improves the separation process [20]. Nuclear density meters are embedded at the end of the deck screen (close to the discharge section) in order to control two main parameters, namely the bed level and the cut height [14]. A rotary star discharge gate is usually utilized for the withdrawal of high-density mineral particles. Since dust generation in dry separation processes is an inevitable issue, a dust collector system is installed to handle the generated dust [14].

The most important benefit of dry jigs is that they completely eliminate the process water, making the air jigs preferable in places with inadequate water or difficult access to it or if the feed materials are sensitive to moisture. Nevertheless, since air density is negligible in comparison to water density; thus, the efficiency of separation during pneumatic jigging is lower than wet jigging. A high-velocity air stream is required to compensate for this significant difference in density, increasing turbulence, and remixing effects. This is why dry jigs are applied only when close-sized material is treated, normally larger than 2 mm. To improve the separation efficiency during dry jigging, it is better to use the input feed with the least near-gravity properties [21].

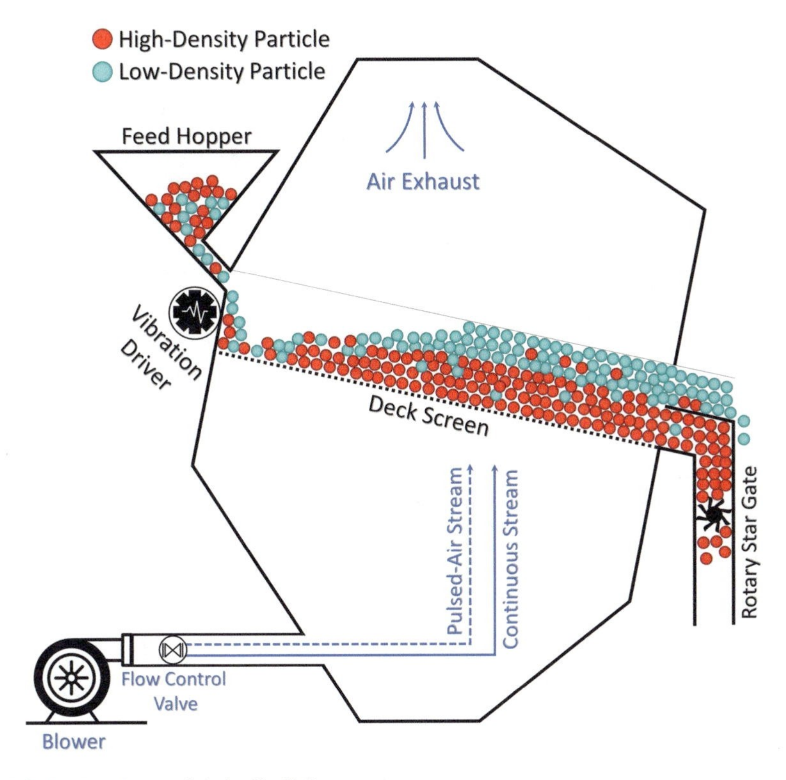

Fig. 3.3 Basic scheme of A dry jig [14]

Because of its specific principal and operational characteristics, air jigs exhibit particular and yet not well-understood stratification dynamics. Despite the wet jigs, there is no suction step in air jigs, and the presence of a continuous air stream results in keeping the bed opened enough to receive periodically pulsed airflow. High-velocity airflow has to be applied to compensate for the effect of low air density. Therefore, uniform and continuous airflow distribution with these specifications is a challenging task, expensive, and requires permanent maintenance. These particular circumstances have prevented the widespread use of dry jigs in mineral processing units [22]. On the other hand, besides the previously mentioned benefits, it has been proved that the stratification process is considerably faster in air jigs because particles can move much more easily in the air rather than in water. This phenomenon can increase the capacity of dry jigging at no additional energy costs [21]. Figure 3.4 demonstrates the jig bed motion and stratification process over a pulsation cycle where: (1) The pulsation velocity is zero, and the bed is static, (2) The pulsation is exerted, and upward drag force raises the entire bed as a rigid body, (3) A high porosity zone is formed below the bed, and the particles start settling, (4) The bed porosity increases and reaches the maximum value, (5) The sedimentation of particles starts based on their gravity, (6) The bed becomes once again static on the jig sieve [14].

Several investigations have been carried out to indicate whether air jigs can be effectively used in mineral processing. For example, in the case of coal beneficiation

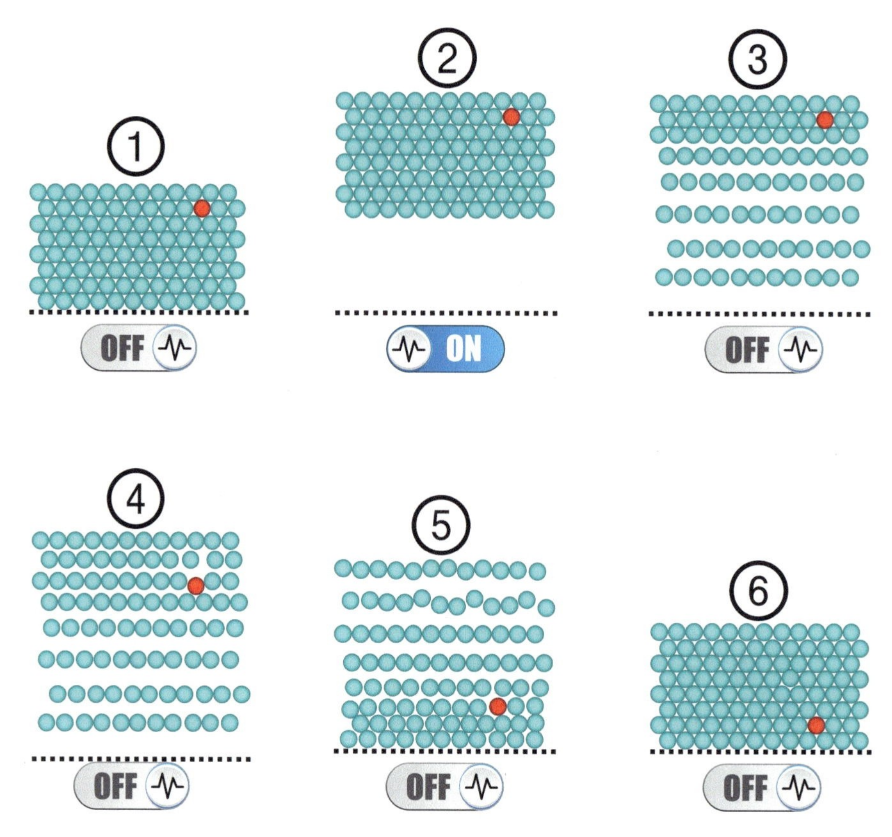

Fig. 3.4 Jig bed motion and stratification process over pulsation cycle [14]

of Candiota mine, the largest Brazilian coal deposit, using air jigging, sulfur content has been significantly reduced (50% of the feed). Thus, the performance of the next step, electrostatic separation, is expected to improve due to the reduction in ash content, leading to lower emission of fly ashes. Consequently, high added-value products will be produced, and purified coal's high calorific value will be obtained [23]. Additionally, the results of beneficiation tests on high-ash, Indian non-coking coal by dry jigging had shown promising results when particles with 5–40 mm were fed to the system. By using dry jigging, the ash content of the coal was reduced by around 7% (from 39.6 to 32.7%). From the ash balance, it was found that the clean coal yield was 75.8% [24].

3.3 Air Table

Air table, also known as pneumatic table, has been originally developed to separate seeds in the food industry. However, many investigations have been carried out to evaluate its performance in mineral processing for coal beneficiation [25–28], sand deposit treatment [29, 30], and upgrading of tungsten [31]. This technology has also received attention in recycling and separating a wide range of secondary materials, including scrap glass, abrasive grains, metal from crushed industrial crucibles, scrap wire from its insulator, and lead from plastics in the spent batteries [32, 33]. Using a pneumatic table has advantages over water-based separation methods because this technology does not need chemical pretreatment. One of the main benefits of applying air tabling is that air keeps the particle bed loose, allowing the particles to move easily to their respective position. However, wet tabling using a shaking table makes the bed compact, preventing the required freedom of movement [34].

Due to its appearance (Fig. 3.5), this machine is termed air table [32]. An Air table principally functions similar to an air jig with a continuous upward fluidizing air stream. This density-based separator mainly includes a feed hopper followed by a vibrating feeder, a porous deck surface connected to a vibrator to apply the eccentric longitudinal vibration, and an electric blower embedded below the deck to create the upward airflow at a controlled volume and superficial velocity. A collecting part is located together with the discharge end of the machine. This part generally comprises two compartments separated by a splitter to divide particles into low-density and high-density fractions. A pyramid-shaped hood has been located above the porous deck

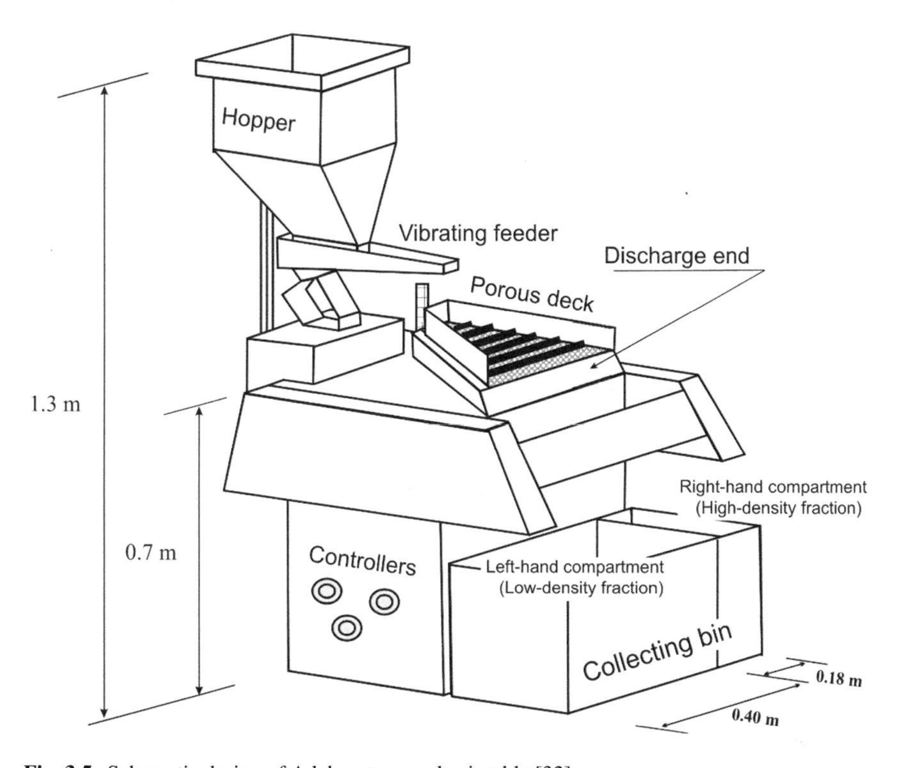

Fig. 3.5 Schematic design of A laboratory-scale air table [32]

surface. This part is occasionally employed to prevent the initial airflow reduction, which may occur close to the sides of the porous deck. The trapezoidal-shaped deck, where the particles move, is made of porous material. This flat porous surface is connected to a perforated plate with circular orifices (approximately 12% opening area) and with a series of parallel-arranged riffles to the direction of longitudinal vibration [28, 34, 35].

The size of the orifices has to be smaller than the size of the smallest particle present in the input feed to prevent the particle from falling into the deck. The slope of the porous deck can be adjusted both longitudinally and transversely, meaning that the plate can become inclined both from side to side and from inlet end to discharge end. This leads to a change in end slope (a) and the side slope (b), respectively. Some controllers are embedded adjacent to the main deck, which permits adjusting operation parameters, including side slope, end slope, airflow superficial velocity, and longitudinal vibration frequency (Fig. 3.6) [32, 34, 35].

The working principles of the air table closely resemble the wet table. Firstly, relatively same-sized particles are fed onto the porous deck from the hopper by the vibrating feeder. This feeding system creates a uniform bed over the deck surface. The eccentric vibrator moves the main deck in a side-to-side direction, along the direction of riffles. The movement frequency (f) can be at the range of 0.09–13.33 s^{-1}. Simultaneously, the electric blower generates an upward airflow through the porous deck, whose velocity can vary from 1 to 4 m s^{-1}. In conjunction with the longitudinal vibration, airflow causes the spreading and lifting of the particle bed on the porous deck surface. Subsequently, as the bed falls, it is expanded and fluidized, resulting in becoming materials stratified based on their density. The high-density materials settle on the deck surface and contact, whereas the low-density materials float on top

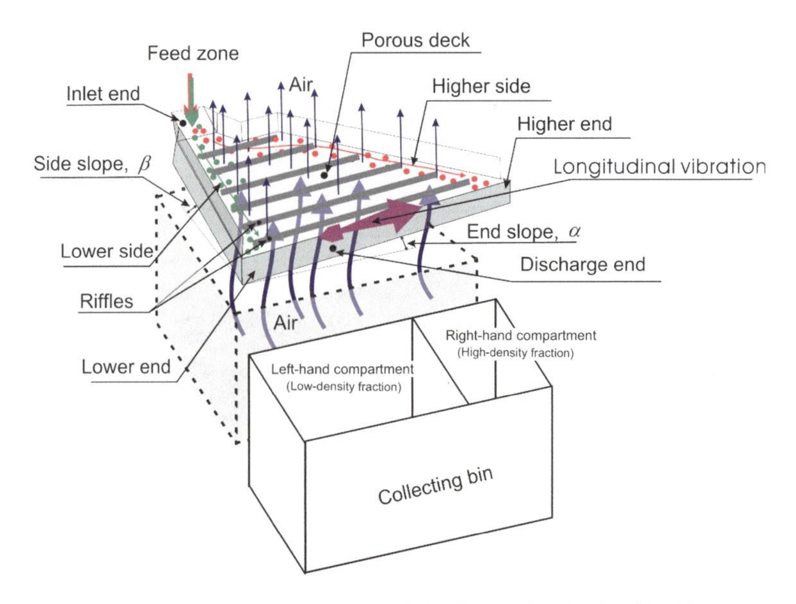

Fig. 3.6 Schematic diagram illustrating the principle of separation by air table [32]

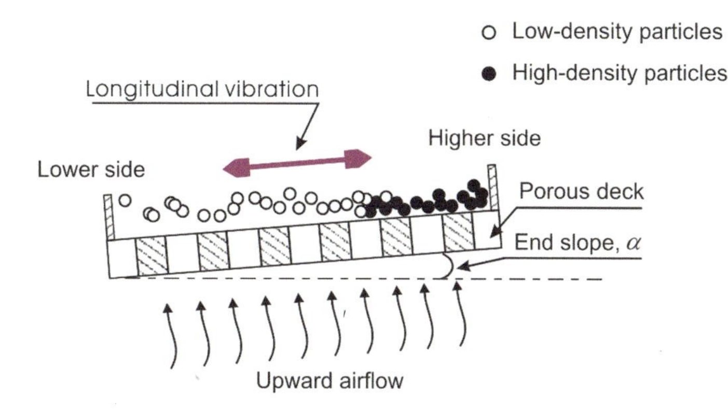

Fig. 3.7 Arrangement of particles with different densities on porous deck [32]

of the high-density particles (Fig. 3.7). The eccentric vibrator moves the porous deck using a slow forward stroke and a quick return. This pattern moves heavier particles along the deck between the riffles [36, 37].

Then, heavy particles flow off the deck through the higher side, which channels them downward to the discharge end, and then drop into the collecting bin. On the other hand, the light particles, which remain fluidized, drift downhill in the direction of the deck's inclination due to gravitational pull and are discharged from the deck at its lower end. The low-density fraction is then collected in another compartment of the collecting bin [35]. The particle motion on the deck is dependent primarily on the deck acceleration. However, the analysis of the material moving along the vibrating deck is not simple since it is influenced by several factors related to the material characteristics and the deck movement [38]. Figure 3.8 shows different forces affecting a particle placed on an inclined porous deck of an air table [32], where:

Q_v Reaction Force
F_{mg} Force of gravity
R Lift Force
F_f Frictional force
φ Angle to horizontal line.

As mentioned earlier, air tabling has been tested in several investigations to evaluate its performance and efficiency. During a set of experiments for beneficiation of fine coal using the air table, it has been indicated that the air table is able to decrease ash content from 27% to approximately 10% at a yield of 80%, which is relatively similar to the results could be provided in the wet processes. It was reported that lower vibration frequencies could provide more efficient separation. This could be because a significant portion of the particles moves toward the tailings discharge at high frequencies, which negatively affects the separation performance. It also has been observed that for particles that are very similar in size, the ash rejection of 80% with a combustible recovery of about 95% can be achieved, indicating a significant separation efficiency. Moreover, pyritic sulfur decreased around 43.3% [28].

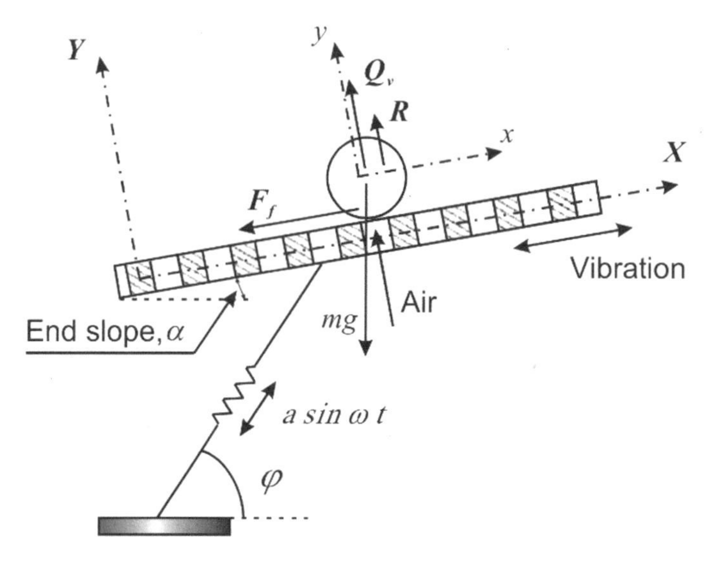

Fig. 3.8 Forces acting on a particle placed on a porous deck during air tabling [35]

3.4 Air-Dense Medium Fluidized Bed Separator

Fluidization can be defined as a process in which mineral particles are suspended by an upward force, flowing gas, for instance, and move like a fluid. Applying this technique in mineral processing was industrially started in 1942 when it was used in catalytic cracking. Thereafter, this concept has been extensively considered in many investigations to understand its principles [39] completely. In the fluidization process, fluidizing medium and its characteristics play an important role. In an air-dense medium fluidized bed separator (ADMFBS), a fluidizing bed with a specific density is created using air along with specific medium particles. In other words, the air causes the particles to form a pseudo-fluid bed in which heavier materials move downward (sink) while high-density particles move upward (float) [40]. For using dry-based fluidized bed process, many investigations have been conducted to explore different alternative materials (magnetite powder, magnetic pearls, a mixture of fine coal and magnetite, and a mixture of hematite and limestone) to utilize as a fluidizing medium to do a separation by using a dry fluidized bed. Magnetite powder has usually been the most suitable option for coal beneficiation among all the materials mentioned. [39].

The fluidization process is defined based on the Archimedes Principle, which states that any object immersed in a liquid, whether completely or partially, is buoyed up by force equal to the weight of the fluid displaced by the object. During this process, particles denser than the fluid move toward the bed bottom (sink), and the lighter particles move up (float) [41]. However, experiments have indicated that some misplacement regarding high-density and low-density particles occurs throughout fluidization, which the principle cannot describe. This phenomenon is called the displaced distribution effect [13, 42]. There are variations to this effect which are known as the displaced movement effect and displaced viscosity effect. The former increases by decreasing airflow, while the latter is influenced when the feed air velocity increases or decreases [42]. The falling sphere model can describe the rheological properties of particles that move in a fluidized bed. Experimental results indicated that the particles in the fluidization process behave similarly to a Bingham fluid. A linear regression of the measured terminal settling velocity can be used to obtain the plastic viscosity and yield stress, which was obtained experimentally. It should be noted that both of the mentioned properties increase with increasing particle feed size [42]. When the diameter of mineral particles is much greater than dense medium particles, similar to what happens in coal beneficiation, the frictional drag force of the air on the coal particles (F_{gd}) can be neglected [43]. In general, four primary forces influence the particle in a fluidized bed, including the gravitational force of the particle (G), friction drag force of the air (F_{gd}), drag force of the air dense medium (F_{sd}), and effective buoyancy force (F_b) (Fig. 3.9) [43].

The hydrodynamics of a fluidization process is complicated and must be understood entirely to enhance the performance efficiency of the separator. It depends on

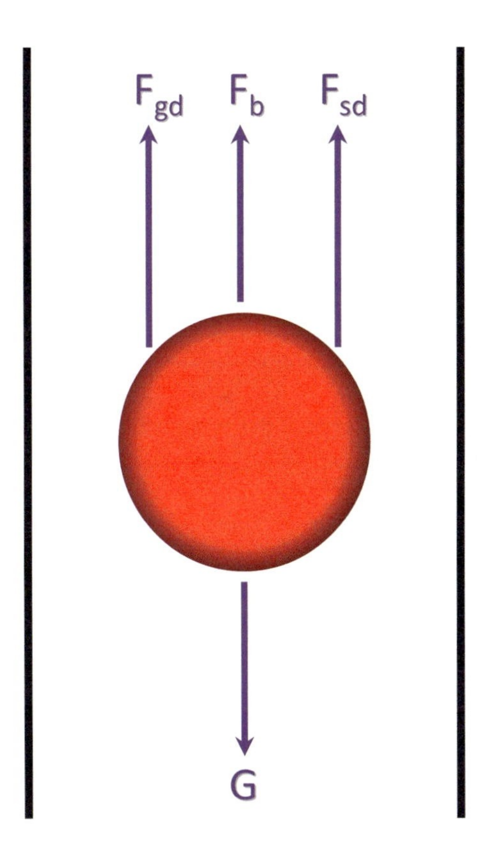

Fig. 3.9 Forces acting on A particle in a fluidized bed [43]

several parameters such as mineral characteristics, air distributor features, gas properties, and other factors that affect the forces between particles and dense medium [40]. Additionally, it is not unexpected that these independent operating parameters interact with each other during fluidization [44]. One of the most important parameters in the fluidization process is minimum fluidization velocity (v_{mf}). It is the velocity of upward airflow at which the drag force and the upward buoyant force due to the fluid are balanced by the particles' weight [43]. This parameter is proportional to the drag force needed to deliver a suspension of the solid particles and is affected by particle properties, fluid properties, and bed geometry [40]. When a separator device using fluidized bed technology is designed, these parameters have to be carefully considered. It is emphasized that an incorrect air velocity in the fluidization process can reduce the separation efficiency. This misplacing effect occurs regarding both viscosity and particle motion [13]. Consequently, since complete fluidization occurs at the minimum fluidization point, precise minimum fluidization velocity is of utmost importance for a successful and effective operation [13, 40].

As shown in Fig. 3.10, the transition of a fixed bed to a fluidized bed occurs when the bed pressure is slightly greater than the static pressure of the bed. This critical pressure is defined as the minimum fluidization point. When the pressure reaches this point, air bubbles start to form if the velocity of the gas is slightly increased. Thus, in practice, when this technology is considered for separation, to achieve complete fluidization, it is necessary to apply air velocity a little greater than the minimum velocity at which the bed starts fluidizing [43, 45].

Air bubbles play an important role in the degree of the fluidization and separation process. The bubbles are closely related to the velocity at which the bed is fluidized. One of the effects of air bubbles on the fluidization process is to shock the particles present in the fluidized bed when passing through the particles. This phenomenon causes the particles to loosen, which permits more effective separation and greater

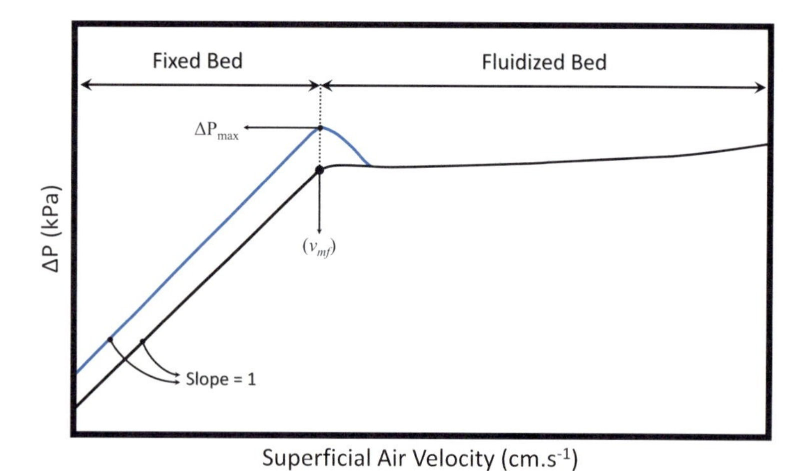

Fig. 3.10 Air velocity versus pressure drop diagram [45]

capacity. However, if the air velocity is not optimal, resulting in undesirable bubbles, it can cause the remixing of different particles and a negative effect on the separation [46]. On the other hand, uniform distribution of air bubbles can cause forming a uniform bed density. This characteristic is necessary to have complete fluidization, and therefore, efficient separation [42]. If large bubbles are formed, disadvantage effects can be expected within the fluidized bed. For example, in coal beneficiation:

- The larger the bubbles, the more likely they are to be burst. Bursting bubbles can intensify the misplacement effect, and thus, fine, high-quality coal particles are misplaced to the tailings section.
- The large bubbles shock the bed to the extent that its height and heterogeneity fluctuate over time and lead to insufficient separation due to inconsistency [41, 46].

Another important parameter regarding the fluidization process is the particle characteristic. In general, different particles can be divided into four groups based on their behavior in the fluidization process:

- Group A: Particles with a low density (lower than 1.4 g cm^{-3}) possessing a small mean particle size. This type of particle can become easily fluidized, and forming controllable bubbles at low gas velocities is expected under this condition.
- Group B: Particles with a density in the range 1.4–4 g cm^{-3} and size in the range 40–500 μm. During their fluidization, the bed group fluidizes with vigorous bubbling. Under this condition, by increasing the bed height, the bubble size also increases.
- Group C: This particle group consists of ultrafine particles. It is very difficult to have a fluidized bed since the forces between particles are greater than the gas force applied to the particles. These are making fluidization almost impossible.
- Group D: Large particles with high density fall into this group. This type of particle makes the fluidization process very difficult while forming and then exploding large bubbles. Furthermore, due to the presence of high-density particles, channeling in the bed and back mixing can also be occurred [40, 43].

The first application of the fluidized bed separation for coal beneficiation was reported in 1926. In that process, river sand with a bulk density of 1.45 g/cm^3 was utilized to clean coal samples in the size range of 10–50 mm. However, the separation performance was never tested on a pilot or industrial scale [39]. Subsequently, dry coal beneficiation was introduced and developed using only gas-solid fluidized beds and coal ranging from 6 to 50 mm [12]. After achieving effective separation during experiments, the first dry coal beneficiation industrial unit was designed and established in China with an annual output of 320,000 tons [42].

Further studies for improving fluidized bed technology for coal beneficiation result in developing the following improved devices to separate a wider size range of coal particles:

- *Duel Density Fluidized Bed*: This technology can divide input feed into three output streams, namely clean, middling, and tailing. For this purpose, this device uses two relatively stable layers of beds with different densities by using two different density materials. Duel-density fluidized beds can replace two fluidized beds working in series and reduce the complexity of the process.

- *Vibrated Air Dense Medium Fluidized Bed*: In this technology, external vibrators supply vibration energy into a conventional fluidized bed to improve the quality of fluidization, and therefore, the separation efficiency. The vibration causes forming of micro-bubbles, which leads to decreasing the back mixing of medium solids. Additionally, employing vibration energy can prevent the fluidizing medium from aggregating and channeling. This happens with supplying the required energy to overcome interparticle forces. Amplification of the contact between particles and air is another effect of vibratory force. According to mentioned features, this method can significantly improve the separation of fine coal.
- *Magnetically Stabilized Fluidized Bed*: This technology exposes the magnetizable fluidizing medium particles to a magnetic field, which is time-invariant and spatially uniform. This process causes the fluidizing medium particles to be axially oriented. Consequently, thanks to the applied magnetic field, the fluidizing bed is stabilized against the growth of disorders, such as bubble growth in the bed. Magnetically stabilized fluidized bed reduces the back mixing effect and facilitates the attainment more stable fluidized bed [11, 39, 42, 47–49].

In separation using the fluidization process, forming uniform microbubbles and stable dispersion fluidization must occur. Furthermore, the three-dimensional bed densities should remain stable during a separation, and the bed is better to have high fluidity and low viscosity [39]. In addition to no need for water and the possibility of use in arid areas, the following advantages can be listed when using dry fluidization for coal beneficiation:

- *High precision*: dry fluidization can effectively separate coal particles. The separation efficiency can be compared with the conventional heavy medium wet beneficiation methods, specifically for coal with a 6–50 mm size range.
- *No environmental pollution*: This environmentally friendly method only requires low-pressure compressed air. Additionally, the devices operate smoothly with very little noise pollution, and the emitted dust by the equipment is within environmental laws.
- *Low investment*: For a specific capacity, a dry fluidization-based beneficiation unit can be designed and established for half the costs compared to a wet beneficiation plant. The main reason is that no expensive slurry treatment is required in the dry method.
- *Wide ranges of densities*: By adding magnetite powder to the bed, beneficiating densities ranging from 1.3 to 2.2 $g.cm^{-3}$ can be achieved. Therefore, this technology can either remove heavy gangues or lower density clean coal depending on the required product [11, 41] Fig. 3.11 illustrates a schematic diagram of an air dense medium fluidized bed separator [13].

Even though there are many advantages to the fluidized bed technology, it has some drawbacks, which include:

- The air-feed is required to become completely moisture-free.

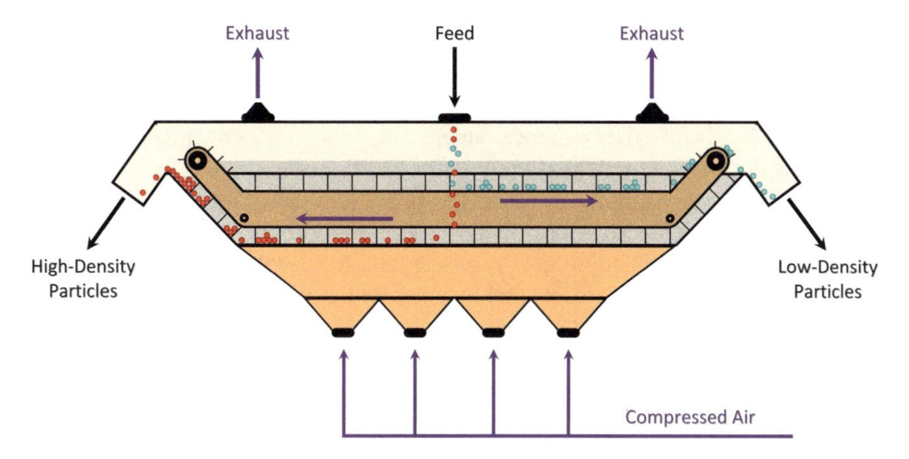

Fig. 3.11 Schematic diagram of air dense medium fluidized bed separator [13]

- The consumption rate of the fluidizing medium is relatively high.
- During the fluidization process, fine coal particles are continuously generated, negatively affecting the overall density of the bed, and therefore, separation efficiency.
- The efficiency of separation significantly decreases when the moisture content of the run of mine (ROM) coal is too high because of reduced fluidity.
- If the input feed size increases, the separation efficiency decreases [13, 50–52].

3.5 Knelson Concentrator

The use of an intensified centrifugal acceleration, which generates an artificial gravity field, has facilitated separating fine particles. The earliest centrifugal device in the mineral processing history was a centrifugal pump used in a copper mine in Portugal in the fifth century [53]. Employing centrifugal separators increases significantly settling velocity, leading to decreasing settling time of particles. This results in smaller separator dimensions, at the same capacity, compared to conventional gravity separators [54]. The first centrifugal concentrator was the Hendy concentrator, which was patented in 1868 and initially employed in California [55]. After that, with the rapid development of technology, many new centrifugal concentrators were developed. The advent of electronic technology in conjunction with new materials made it possible to develop improved centrifugal concentrators with advantageous features such as high centrifugal acceleration, resulting in high capacity and better performance [54].

The Knelson concentrator, an improved density-based separator, is principally a centrifugal concentrator that uses a fluidized bed to concentrate the fine size material [56]. This machine is extensively utilized in separating gold and other precious metals. Knelson concentrator was developed in 1978 and commercialized in 1980 in Canada by Byron Knelson [57]. As Fig. 3.12 shows, the design of this vertical axis bowl-type concentrator was initially based on fluidizing water stream. The bowl comprises a conical inner shell, with a series of riffles attached to a revolving outer shell. Input mixture, most frequently ball mill discharge or cyclone underflow, is fed through a central tube as slurry into the bowl. The bowl rotates rapidly, which generates an artificially boosted gravity field. This procedure considerably increases sedimentation velocity differential. As a result, improved centrifugal gravity concentrators can separate fine particles with high efficiency [56, 57].

An approximately 60–100 G centrifugal acceleration causes the input particles to fill the inter-riffle spaces from bottom to top. When the riffles become full of particles, the sorting begins, where high-density materials displace the light particles. This process causes high-density particles to be trapped in the inter-riffle spaces, whereas the water stream moves low-density materials to the top of the unit and ejects them as tailings. The separation occurs through bed fluidization in the riffle, permitting the replacement of denser particles with light materials. The fluidization is generated by water injected through orifices in the riffles. This fluidization has to be strong enough

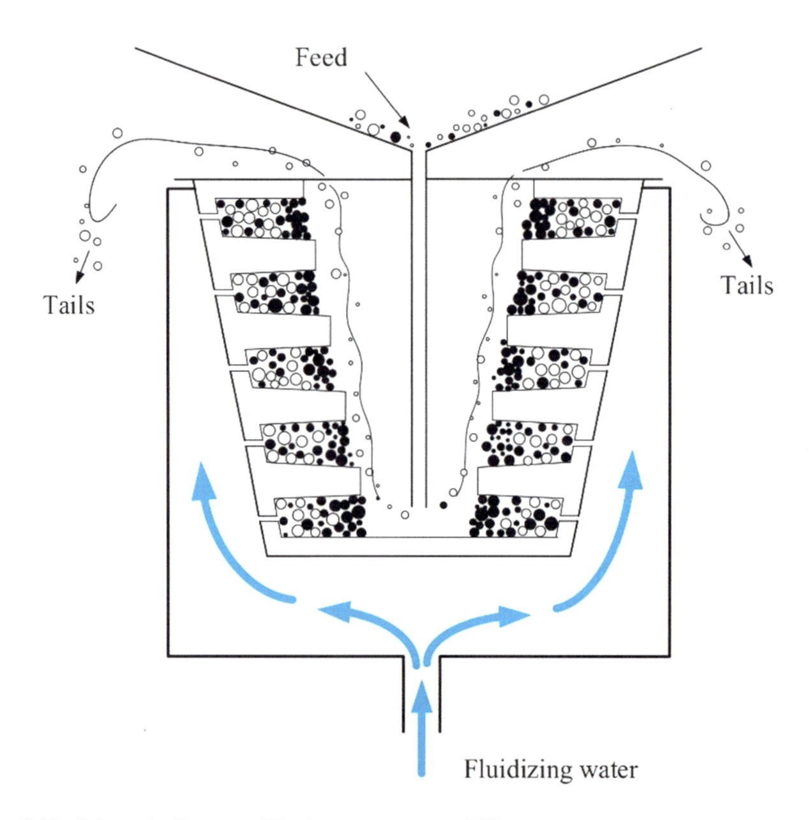

Fig. 3.12 Schematic diagram of Knelson concentrator [57]

to prevent compaction of the dense particle bed because of the strong centrifugal force [57, 58].

Although centrifugal-based machines have achieved a high-quality final product with excellent recovery and high separation efficiency, they also have some drawbacks. One of the disadvantages of these separators is that they usually operate on a wet basis, and a large amount of water is needed during the separation process. For instance, at the laboratory scale experiment, for processing around 24 kg of ore, approximately 300 L of water are required [59]. Additionally, Since centrifugal separation is usually utilized for very fine particles, in addition to water consumption and wastewater treatment, filtration and thickener costs are also increased during wet beneficiation. Therefore, there are growing pressures to minimize water consumption, and even dry processing is considered [60].

It newly has been indicated that Knelson Concentrator has great potential for dry processing. In an experimental study, to separate tungsten from quartz using a Knelson Concentrator, the air was utilized as the fluidizing medium in a synthetic ore (1% w/w tungsten). The obtained results have shown that a tungsten concentrate of 6.32% tungsten grade can be produced with around 78.5% recovery, indicating the possibility of operating a Knelson Concentrator on a dry basis [61]. Another investigation has reported that in separating tungsten from quartz using dry Knelson Concentrator, the most important parameters are particle size and solid feed rate. Based on the experimental tests, the recovery of 30, 50, and 65% for tungsten can be achieved regarding coarse ($-425 + 300$ μm), middle ($-150 + 106$ μm) and fine (-53 μm) fraction sizes, respectively. [62]. Figure 3.13 shows a diagram for a dry Knelson Concentrator [63].

3.6 Reflux Classifier

The Reflux Classifier is another machine that was initially developed for wet beneficiation applications. However, due to water scarcity and the advantages of waterless techniques, the possibility of using it in dry mineral processing has recently been examined [64]. As illustrated in Fig. 3.14, the Reflux classifier comprises several parallel inclined channels above a conventional fluidized bed. The inclined channels provide a larger effective segregation area due to the Boycott Effect, allowing much higher operating velocities than conventional fluidized beds [65, 66]. A reflex action occurs because of particle segregation onto the upward inclined surfaces. Subsequently, the settled mineral particles slide down the inclined surfaces of channels and come back to the fluidized zone. The particles are re-fluidized and return to the inclined channels for further possible separation to the overflow. Particles, especially low-density ones, can also experience a mechanism of re-suspension in the channels due to shear-induced lift forces [66, 67]. Since fluidized beds can provide a suitable environment for mixing the particles, their application for separation requires circumstances that minimize the mixing effect. The separation process in a fluidized

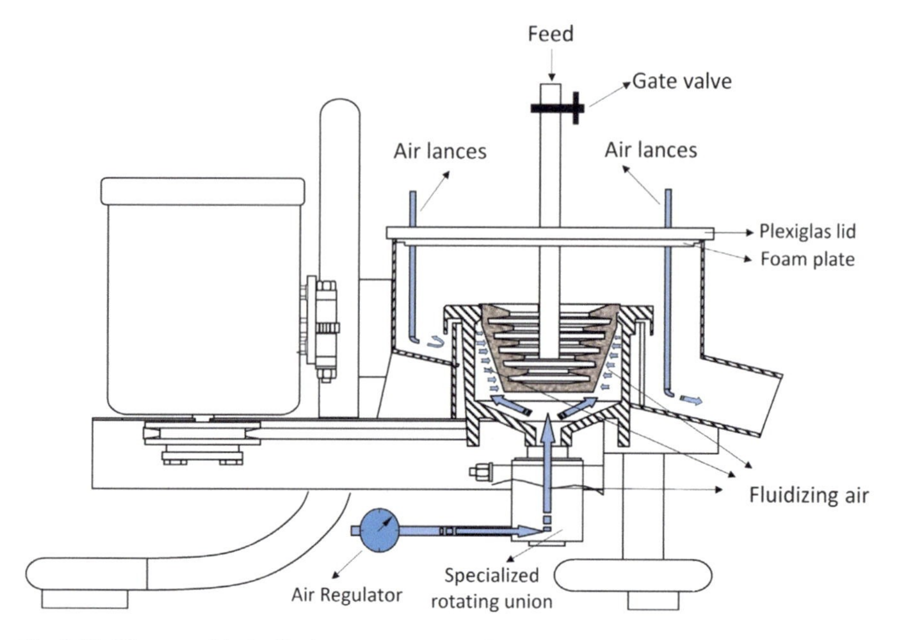

Fig. 3.13 Diagram of A dry Knelson concentrator [63]

bed happens because of the difference in the final settling velocity of the various particles within the bed. In this process, the particles possessing high settling velocity accumulate at the bed bottom, and the lowest settling velocity particles move toward the top of the bed [68, 69].

However, the initial design of this method was based on the use of water as a separation medium; it is possible to use air with a fluidized bed of fine solid particles as an alternative medium. This type of medium possesses an effective density intermediate between the densities of air and constituent solid particles. The quality of fluidization, and consequently, the separation efficiency, wholly depends on the size of the fine particles present in fluidizing medium and their properties. For achieving an efficient separation, the medium density has to be consistent [66, 69]. Vibration decreases the size of bubbles, and therefore, can provide a more consistent density throughout the fluidized bed. Moreover, vibration reduces the minimum fluidization velocity of the particles, which facilitates the fluidization process [70].

In the Reflux Classifying, vibration may also contribute to the re-suspension of particles in the inclined settling zone. It is reported that fluidization using gases can be highly effective for density-based coal beneficiation [71]. The first investigation on the application of a dry Reflux Classifier was conducted in 2007. It has been reported that the effect of parallel inclined channels on the elutriation of particles from the dilute phase freeboard of an air-fluidized bed is very similar to the water-based system. It has also been confirmed that the inclined channels allow higher operating velocities for the same separation size, leading to more stable and sharper separations [72]. Further research has also been conducted to investigate the effect of operating parameters, such as particle size, underflow rate, and gas flow rate, on

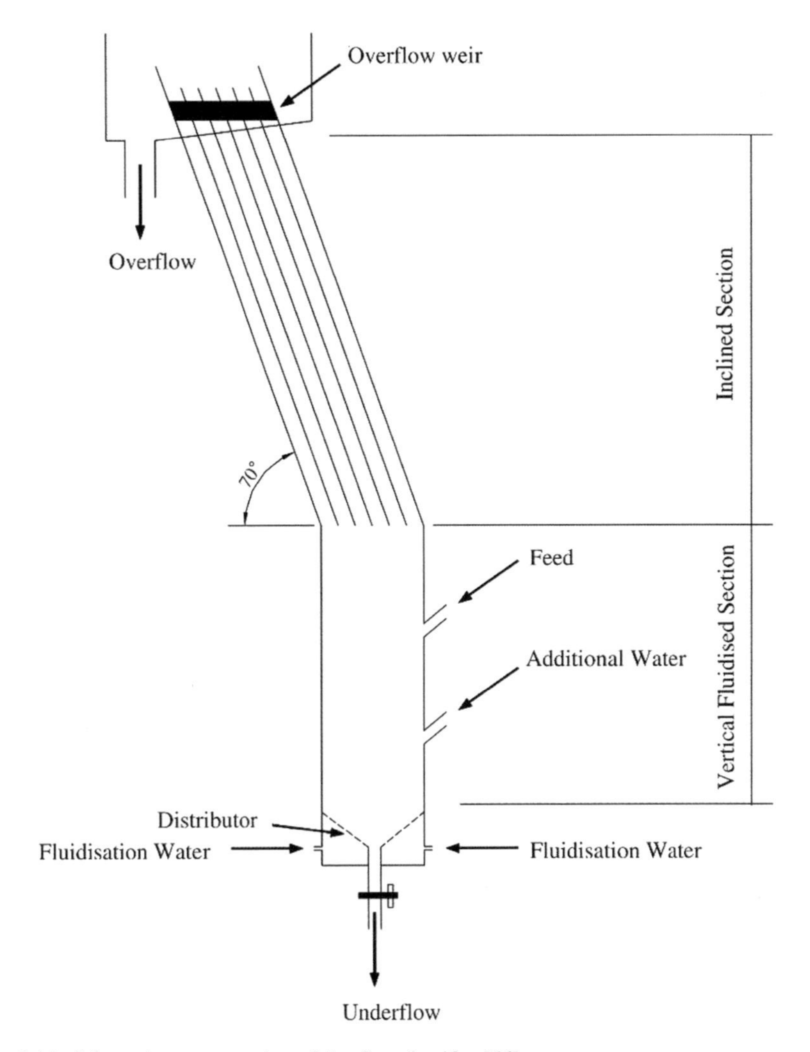

Fig. 3.14 Schematic representation of A reflux classifier [66]

the density cut point and separation efficiency. The results obtained have shown that separation efficiency decreases when particle size increases. Also, increasing the overflow or gas rates increased the density separation cut point [64, 73].

3.7 Main Applications and Producers

These methods and systems have been used in different mineral processing investigations, summarized in Table 3.1. The representative producers of dry gravity separators are presented in Table 3.2.

Table 3.1 Summary of the dry gravity separator's application for various mineral beneficiation

Separator	Feed	Mine location	Feed size	Variable studied	Remarks	References
Air table	Coal	India	0–1 mm	– Feed rate – Table side tilt – Table end tilt – Table frequency	– Very good rejection of ash is achievable using an air table with a nearly 10% reduction in the ash level of the clean coal in a single-stage operation	[25]
Air table	Coal	USA	0–6.3 mm	– Feed size – Table frequency – Longitudinal angle – Transverse angle	– For the 3.35–6.3 mm size coal fraction, the table frequency, longitudinal angle, and transverse angle were found to be important variables. Whereas, for the 1.4–3.35 mm size fractions, only table frequency, and transverse angles were found to be important	[28]
Air table	Coal	Turkey	6–38 mm	– Moisture content – Feed rate	– Moisture content had a remarkable effect on separation efficiency, and it was inversely affected when the moisture content increased	[74]
Air table	Coal	South Korea	1–20 mm	– Feed rate – Bed height – Feed type: the ratio of light and dense fractions in the feed	– In order to improve the vertical stratification process in the bed, it was required to increase the bed height on the air table during the separation process	[27]

(continued)

Table 3.1 (continued)

Separator	Feed	Mine location	Feed size	Variable studied	Remarks	References
Air table	low-rank Coal	USA	0–75 mm	– Feed size – Blower frequency – Table frequency – Longitudinal angle – Transverse angle	– Sufficient upgrading can be achieved on the +1 mm fraction; however, the size fraction −6 mm has a greater tendency to report to the product stream, and hence, It would require a different process to clean	[75]
Air table	Coal	Turkey	0–40 mm	– Table Frequency – Table slope – Feed Rate – Riffle height	– Increasing feed rate and table slope had negative effects on calorific values and ash content of clean coal. On the other hand, increasing table frequency and riffle height had positive effects on calorific values and the tailing ash content of clean coal. However, it was found that the most effective parameter is particle size distribution – The most favorable feed size for this apparatus has been between 6 and 38 mm	[26]

(continued)

Table 3.1 (continued)

Separator	Feed	Mine location	Feed size	Variable studied	Remarks	References
Air table	Coal	Mongolia	1–25 mm	– Feed rate – Airflow frequency – Table frequency – Cross table slop - Longitudinal slope	– The size fraction of 1–5 mm showed that the air table is effectively removing high-density reject materials from feed coal – Effective feeding rate is 4.5–5 ton-h^{-1} for 1–5 mm size fraction and 20–30 ton-h^{-1} for 5–10 mm coal	[76]
Air table	Coal	India	0–1.78 mm	– Table frequency – Table side tilt – Table end tilt – Air velocity	– Two-stage processing was found to be highly effective for improving the combustibles in the clean coal – The ash level could be reduced significantly from 55 to 35%	[36]
Dry jig	Coal	India	5–40 mm	–	– Dry jigging of a high ash feed with an ash content of 40% was successful – A 50 t/h capacity jig was installed and commissioned to beneficiate high-ash coals – A clean coal product of 33% ash was produced and is being used for a sponge iron plant	[24]

(continued)

Table 3.1 (continued)

Separator	Feed	Mine location	Feed size	Variable studied	Remarks	References
Dry jig	Coal	Brazil	0–50.8 mm	–	– Feed with 1.8% sulfur and about 51% ash appeared after jigging with about 0.7% sulfur and near 47% ash. After these results, pilot jigs are planned to install at the plant	[23]
Dry jig	Coal	Mozambique	1–50 mm	– Feed type: Coal from the various layer of Moatize coal have been tested	– A significant difference was shown in the coal mass recovery profile from distinct layers; however, dry jigging can remove about 18% of all heavies (>2.2 g cm^{-3}) of the Moatize coal seam	[77]
ADMFB	Coal	India	13–50 mm Medium: Magnetite powder 63–150 μm	– Feed size – Medium size	– ADMFB is not suitable for beneficiation of examined coal of 1–13 mm size fraction since the separation efficiency decreases sharply at coal particle sizes less than 13 mm and the process becomes uneconomical – The superior performance of the ADMFB separator occurred when 106–150 μm magnetite powder was used as the fluidizing medium	[78]

(continued)

Table 3.1 (continued)

Separator	Feed	Mine location	Feed size	Variable studied	Remarks	References
ADMFB	Bauxite	China	6–18 mm Medium: Fe-Si powder 74–425 μm	– Medium size – Separating time – Static bed heights – Gas velocities	– The fraction size of 125–180 μm for Fe-Si powder was appropriate for the fluidization stability, and thus, it can be utilized as the dense fluidizing medium for separation – Al/Si ratios were effectively improved, and the product can be used for the subsequent extraction of Aluminum metal using the Bayer process	[79]
ADMFB	Coal	Australia	5–31.5 mm Medium: Silica-Zircon sand	– Particle size – Bed height	– Particle size and bed height were found to strongly influence the efficiency of separation by the dry dense-medium separator – Satisfactory results of the separation tests indicate that the dry sand fluidized bed dense-medium separator could be an alternative process for efficient coal beneficiation without the use of water	[80]

(continued)

Table 3.1 (continued)

Separator	Feed	Mine location	Feed size	Variable studied	Remarks	References
ADMFB	Iron ore	China	6–31.5 mm Medium: Atomized iron powder and zircon sand 75–250 μm	– Feed size – Static bed heights – Gas velocities – Medium proportion	– The stability of bed density was maintained under the condition with a static bed height of 80–90 mm, an operational gas velocity of 1.6–2.0Umf, and a weight the proportion of the atomized iron powder in the binary dense media of 90% – The separation efficiency of coarse iron ore particles gradually decreased with decreasing size fraction	[81]
ADMFB	Copper ore	Chile	10–25 mm	– Medium type – Air velocity	– Copper ores can be efficiently separated at target set points within the range of about 2200 to 3700 kg m^{-3} – 44 to 48% of the lowest density ore could be removed with little loss of copper	[82]
ADMFB	Coal	China	6–50 mm Medium: magnetite powder and the mixture of magnetite powder and fine coal 0–500 μm	– Medium type	– ADMFB separator can separate effectively 6–50 mm coal – Both fluidizing medium types can be used effectively to separate the feed	[41]
Knelson	Tungsten-Quartz	synthetic	38–425 μm	– Bowl speed – Solid feed rate – Air pressure	– Bowl speed and air fluidizing pressure were significant for both the grade and recovery	[63]

(continued)

Table 3.1 (continued)

Separator	Feed	Mine location	Feed size	Variable studied	Remarks	References
Knelson	Tungsten-Quartz	synthetic	38–425 μm	– Feed size – Feed rate – Motor power – Air pressure	– The motor power had the greatest impact when comparing to air fluidizing pressure and solid feed rate – Different optimum conditions had been found for different size fractions	[62]

Table 3.2 The main dry gravity separator producers

Equipment	Manufacturer	Location
Gravity separator	B + B Anlagenbau	Germany
	Bollegraaf	Netherlands
	Dieffenbacher	Germany
	Eaton	USA
	Erga	Russia
	Flsmidth	Denmark
	GEA Westfalia	Germany
	GN Solids Control	China
	Herbold	UK
	Huber	Germany
	Longi	China
	Metso	Finland
	Montalbano	Italy
	Salher	India
	Scheuch	Austria
	Spaleck	Germany
	Steinhaus	Germany
	Turatti	Italy
	Westech	USA

References

1. Burt, R.: Role of gravity concentration in modern processing plants. Miner. Eng. **12**, 1291–1300 (1999). https://doi.org/10.1016/S0892-6875(99)00117-X
2. Dodbiba, G., Haruki, N., Shibayama, A., Miyazaki, T., Fujita, T.: Combination of sink–float separation and flotation technique for purification of shredded PET-bottle from PE or PP flakes. Int. J. Miner. Process. **65**, 11–29 (2002). https://doi.org/10.1016/S0301-7516(01)00056-4
3. Truscott, S.J.: A Text-Book of Ore Dressing. Macmillan and Co. Ltd. (1923)
4. Falconer, A.: Gravity separation: old technique/new methods. Phys. Sep. Sci. Eng. **12**, 31–48 (2003). https://doi.org/10.1080/1478647031000104293
5. Kawatra, S.K.: Advanced coal preparation and beyond: CO_2 capture and utilization. Adv. Coal Prep. Beyond (2020). https://doi.org/10.1201/9780429288326
6. Wills' Mineral Processing Technology. Elsevier (2016)
7. Holland-Batt, A.B.: Gravity separation: a revitalized technology. Min. Eng. **50**, 43–48 (1998)
8. Subba Rao, D.V.: Mineral Beneficiation a Concise Basic Course (2011). ISBN 9780203847893
9. Honaker, R.Q., Forrest, W.R.: Advances in gravity concentration. 190 (2003)
10. Dwari, R.K., Rao, K.H.: Dry beneficiation of coal—a review. Miner. Process. Extr. Metall. Rev. **28**, 177–234 (2007). https://doi.org/10.1080/08827500601141271
11. Chen, Q., Wei, L.: Coal dry beneficiation technology in china: the state-of-the-art. China Particuol. **1**, 52–56 (2003). https://doi.org/10.1016/s1672-2515(07)60108-0
12. Zhao, Y.M., Li, G.M., Luo, Z.F., Liang, C.C., Tang, L.G., Chen, Z.Q., Xing, H.B.: Modularized dry coal beneficiation technique based on gas-solid fluidized bed. J. Cent. South Univ. **18**, 374–380 (2011). https://doi.org/10.1007/s11771-011-0706-6

13. Sahu, A.K., Biswal, S.K., Parida, A.: Development of air dense medium fluidized bed technology for dry beneficiation of coal—a review. Int. J. Coal Prep. Util. **29**, 216–241 (2009). https://doi.org/10.1080/19392690903113847
14. Ambrós, W.M.: Jigging: a review of fundamentals and future directions. Minerals **10**, 1–29 (2020). https://doi.org/10.3390/min10110998
15. Boylu, F., Tali, E., Çetinel, T., Çelik, M.S.: Effect of fluidizing characteristics on upgrading of lignitic coals in gravity based air jig. Int. J. Miner. Process. **129**, 27–35 (2014). https://doi.org/10.1016/J.MINPRO.2014.04.001
16. Boylu, F., Tasdemiroglu, E.V., Karagaçlioglu, I., Çetinel, T., Çinku, K.: Dry processing of leonardite by air based gravity separators. In: XII International Mineral Processes Symposium, pp. 10–12 (2012)
17. Weitkaemper, L., Wotruba, H.: Effective dry density benefication of coal. In: XXV International Mineral Processing Congress 2010, IMPC 2010, vol. 5, pp. 3687–3693 (2010)
18. Wotruba, H., Weitkaemper, L., Steinberg, M.: Development of a new dry density separator for fine-grained materials. In: XXV International Mineral Processing Congress 2010, IMPC 2010, vol. 2, pp. 1393–1398 (2010)
19. Snoby, R.W.J.: Advances in dry jigging improves coal quality. Min. Eng. 29–34 (2007)
20. Snoby, R., Thompson, K., Mishra, S., Snoby, B.: Dry jigging coal: case history performance. SME Annual Meeting, pp. 9–52
21. Sampaio, C.H., Cazacliu, B.G., Miltzarek, G.L., Huchet, F., Le Guen, L., Petter, C.O., Paranhos, R., Ambrós, W.M., Silva Oliveira, M.L.: Stratification in air jigs of concrete/brick/gypsum particles. Constr. Build. Mater. **109**, 63–72 (2016). https://doi.org/10.1016/J.CONBUILDMAT.2016.01.058
22. Coelho, A., De Brito, J.: Economic viability analysis of a construction and demolition waste recycling plant in Portugal—Part I: location, materials, technology and economic analysis. J. Clean. Prod. **39**, 338–352 (2013). https://doi.org/10.1016/J.JCLEPRO.2012.08.024
23. Sampaio, C.H., Aliaga, W., Pacheco, E.T., Petter, E., Wotruba, H.: Coal beneficiation of Candiota mine by dry jigging. Fuel Process. Technol. **89**, 198–202 (2008). https://doi.org/10.1016/J.FUPROC.2007.09.004
24. Gouri Charan, T., Chattopadhyay, U.S., Singh, K.M.P., Kabiraj, S.K., Haldar, D.D.: Beneficiaron of high-ash, Indian non-coking coal by dry jigging. Miner. Metall. Process. **28**, 21–23 (2011). https://doi.org/10.1007/BF03402320
25. Chalavadi, G., Singh, R.K., Das, A.: Processing of coal fines using air fluidization in an air table. Int. J. Miner. Process. **149**, 9–17 (2016). https://doi.org/10.1016/j.minpro.2016.02.002
26. Kademli, M., Gulsoy, O.Y.: Investigation of using table type air separators for coal cleaning. Int. J. Coal Prep. Util. **33**, 1–11 (2013). https://doi.org/10.1080/19392699.2012.717566
27. Jambal, D., Kim, B.G., Jeon, H.S.: Dry beneficiation of coal on KAT air table. Int. J. Coal Prep. Util. **00**, 1–13 (2020). https://doi.org/10.1080/19392699.2020.1780428
28. Patil, D.P., Parekh, B.K.: Beneficiation of fine coal using the air table. Int. J. Coal Prep. Util. **31**, 203–222 (2011). https://doi.org/10.1080/19392699.2011.574948
29. Canning, E.: Dry mineral sand separation plants of associated mineral consolidated Ltd. New South Wales Queensland, Min. Met. Pr. Australia, pp. 758–761 (1980)
30. Hudson, S.B.: Air tabling of beach sand non-conductors for zircon recovery. Proc. Aust. Inst. Min. Met. **204**, 81–108 (1962)
31. Osborn, W.X.: Tabling tungsten ore without water. Eng. Min. J **123**, 287–289 (1927)
32. Dodbiba, G., Shibayama, A., Miyazaki, T., Fujita, T.: Separation performance of PVC and PP plastic mixture using air table. Phys. Sep. Sci. Eng. **12**, 71–86 (2003). https://doi.org/10.1080/1478647031000139385
33. Weiss, N.: SME mineral processing handbook. Society of Mining Engineers of the American Institute of Mining Metallurgical and Petroleum Engineers: New York, N.Y. (1985). ISBN 9780895204486
34. Akbari, H., Ackah, L., Mohanty, M.: Performance optimization of a new air table and flip-flow screen for fine particle dry separation. Int. J. Coal Prep. Util. **40**, 581–603 (2020). https://doi.org/10.1080/19392699.2017.1389727

35. Sivamohan, R., Forssberg, E.: Principles of tabling. Int. J. Miner. Process. **15**, 281–295 (1985). https://doi.org/10.1016/0301-7516(85)90046-8
36. Dey, S., Gangadhar, B., Gopalkrishna, S.J.: Amenability to dry processing of high ash thermal coal using a pneumatic table. Int. J. Min. Sci. Technol. **25**, 955–961 (2015). https://doi.org/10.1016/j.ijmst.2015.09.012
37. Patro, S., Chalavadi, G., Singh, R.K.: Residence time distribution studies in air table. Int. J. Coal Prep. Util. **00**, 1–18 (2019). https://doi.org/10.1080/19392699.2019.1699073
38. Colijn, H.: Mechanical conveyors for bulk solids (1985)
39. Mohanta, S., Rao, C.S., Daram, A.B., Chakraborty, S., Meikap, B.C.: Air dense medium fluidized bed for dry beneficiation of coal: technological challenges for future. Part. Sci. Technol. **31**, 16–27 (2013). https://doi.org/10.1080/02726351.2011.629285
40. Escudero, D., Heindel, T.J.: Bed height and material density effects on fluidized bed hydrodynamics. Chem. Eng. Sci. **66**, 3648–3655 (2011). https://doi.org/10.1016/J.CES.2011.04.036
41. Zhenfu, L., Qingru, C.: Dry beneficiation technology of coal with an air dense-medium fluidized bed. Int. J. Miner. Process. **63**, 167–175 (2001). https://doi.org/10.1016/S0301-7516(01)00049-7
42. Chen, Q., Wei, L.: Development of coal dry beneficiation with air-dense medium fluidized bed in china. China Particuol. **3**, 42 (2005). https://doi.org/10.1016/s1672-2515(07)60161-4
43. Kunni, D., Levenspiel, O.: Fluidization Engineering (1991)
44. Mohanta, S., Chakraborty, S., Meikap, B.C.: Influence of coal feed size on the performance of air dense medium fluidized bed separator used for coal beneficiation. Ind. Eng. Chem. Res. **50**, 10865–10871 (2011). https://doi.org/10.1021/ie201548r
45. Chandimal Bandara, J., Sørflaten Eikeland, M., Moldestad, B.M.E.: Analyzing the effects of particle density, size, size distribution and shape for minimum fluidization velocity with Eulerian-Lagrangian CFD simulation. In: Proceedings of 58th Conference on Simulation Modelling (SIMS 58), Reykjavik, Iceland, 25–27 Sept 2017, vol. 138, pp. 60–65. https://doi.org/10.3384/ecp1713860
46. Luo, Z.F., Tang, L.G., Dai, N.N., Zhao, Y.M.: The effect of a secondary gas-distribution layer on the fluidization characteristics of a fluidized bed used for dry coal beneficiation. Int. J. Miner. Process. **118**, 28–33 (2013). https://doi.org/10.1016/j.minpro.2012.12.001
47. Lubin, W., Qingru, C., Zhenfu, L.: Fine coal and three product dry beneficiation with vibration and double-density fluidized beds. J. Cent. South Univ. Technol. (English Ed.) **5**, 106–107 (1998). https://doi.org/10.1007/s11771-998-0047-2
48. Fan, M., Chen, Q., Zhao, Y., Luo, Z.: Fine coal (6–1 mm) separation in magnetically stabilized fluidized beds. Int. J. Miner. Process. **63**, 225–232 (2001). https://doi.org/10.1016/S0301-7516(01)00054-0
49. Yang, X., Zhao, Y., Luo, Z., Song, S., Chen, Z.: Fine coal dry beneficiation using autogenous medium in a vibrated fluidized bed. Int. J. Miner. Process. **125**, 86–91 (2013). https://doi.org/10.1016/j.minpro.2013.10.003
50. Luo, Z., Zhu, J., Fan, M., Zhao, Y., Tao, X.: Low density dry coal beneficiation using an air dense medium fluidized bed. J. China Univ. Min. Technol. **17**, 306–309 (2007). https://doi.org/10.1016/S1006-1266(07)60094-7
51. He, Y., Zhao, Y., Chen, Q.: Fine particle behavior in air fluidized bed dense medium dry separator. Int. J. Coal Prep. Util. **23**, 33–45 (2003). https://doi.org/10.1080/07349340302268
52. Tang, L.: Characteristics of fluidization and dry-beneficiation of a wide-size-range medium-solids fluidized bed. Int. J. Min. Sci. Technol. **27**, 467–471 (2017). https://doi.org/10.1016/j.ijmst.2017.03.013
53. Özgen, S., Arsoy, Z., Ersoy, B., Çiftçi, H.: Coal recovery from coal washing plant tailings with Knelson concentrator. Int. J. Coal Prep. Util. **00**, 1–11 (2019). https://doi.org/10.1080/19392699.2019.1665033
54. Coulter, T., Subasinghe, G.K.N.: A mechanistic approach to modelling Knelson concentrators. Miner. Eng. **18**, 9–17 (2005). https://doi.org/10.1016/j.mineng.2004.06.035
55. Louis, H.: A Handbook of Gold Milling. Macmillan and Co. (1894)

56. Knelson, B., Edwards, R.: Development and Economic Application of Knelson Concentrators in Low Grade Alluvial Gold Deposits. Undefined (1990)
57. Chen, Q., Yang, H., Tong, L., Niu, H., Zhang, F., Chen, G.: Research and application of a Knelson concentrator: a review. Miner. Eng. **152** (2020). https://doi.org/10.1016/j.mineng.2020.106339
58. Knelson, B.: The Knelson concentrator. Metamorphosis from crude beginning to sophisticated world wide acceptance. Miner. Eng. **5**, 1091–1097 (1992). https://doi.org/10.1016/0892-6875(92)90151-X
59. Laplante, A.R., Woodcock, F., Noaparast, M.: Predicting gravity separation gold recoveries. Min. Metall. Explor. **12**(12), 74–79 (1995). https://doi.org/10.1007/BF03403081
60. Oshitani, J., Franks, G.V., Griffin, M.: Dry dense medium separation of iron ore using a gas–solid fluidized bed. Adv. Powder Technol. **21**, 573–577 (2010). https://doi.org/10.1016/J.APT.2010.02.014
61. Greenwood, M., Langlois, R., Waters, K.E.: The potential for dry processing using a Knelson concentrator. Miner. Eng. **45**, 44–46 (2013). https://doi.org/10.1016/j.mineng.2013.01.014
62. Zhou, M., Kökkiliç, O., Langlois, R., Waters, K.E.: Size-by-size analysis of dry gravity separation using a 3-in. Knelson concentrator. Miner. Eng. **91**, 42–54 (2016). https://doi.org/10.1016/j.mineng.2015.10.022
63. Kökkiliç, O., Langlois, R., Waters, K.E.: A design of experiments investigation into dry separation using a Knelson concentrator. Miner. Eng. **72**, 73–86 (2015). https://doi.org/10.1016/j.mineng.2014.09.025
64. MacPherson, S.A., Iveson, S.M., Galvin, K.P.: Density-based separation in a vibrated Reflux Classifier with an air-sand dense-medium: tracer studies with simultaneous underflow and overflow removal. Miner. Eng. **24**, 1046–1052 (2011). https://doi.org/10.1016/j.mineng.2011.05.002
65. Doroodchi, E., Fletcher, D.F., Galvin, K.P.: Influence of inclined plates on the expansion behaviour of particulate suspensions in a liquid fluidised bed. Chem. Eng. Sci. **59**, 3559–3567 (2004). https://doi.org/10.1016/J.CES.2004.05.020
66. Zhou, J., Walton, K., Laskovski, D., Duncan, P., Galvin, K.P.: Enhanced separation of mineral sands using the Reflux Classifier. Miner. Eng. **19**, 1573–1579 (2006). https://doi.org/10.1016/J.MINENG.2006.08.009
67. Galvin, K.P., Doroodchi, E., Callen, A.M., Lambert, N., Pratten, S.J.: Pilot plant trial of the reflux classifier. Miner. Eng. **15**, 19–25 (2002). https://doi.org/10.1016/S0892-6875(01)00193-5
68. Rhodes, M.: Introduction to Particle Technology, 2nd edn., pp. 1–450 (2008). https://doi.org/10.1002/9780470727102
69. Jin, H., Tong, Z., Schlaberg, H.I., Zhang, J.: Separation of fine binary mixtures under vibration in a gas-solid fluidized bed with dense medium. Waste Manage. Res. **23**, 534–540 (2005). https://doi.org/10.1177/0734242X05060858
70. Wang, T., Jin, Y., Wang, Z., Yu, Z.: The characteristics of wave propagation in a vibrating fluidized bed. Chem. Eng. Technol. **20**, 606–611 (1997). https://doi.org/10.1002/CEAT.270200906
71. Lockhart, N.C.: Dry beneficiation of coal. Powder Technol. **40**, 17–42 (1984). https://doi.org/10.1016/0032-5910(84)85053-6
72. Callen, A., Moghtaderi, B., Galvin, K.P.: Use of parallel inclined plates to control elutriation from a gas fluidized bed. Chem. Eng. Sci. **1–2**, 356–370 (2007). https://doi.org/10.1016/J.CES.2006.08.057
73. MacPherson, S.A., Galvin, K.P.: The effect of vibration on dry coal beneficiation in the reflux classifier. Int. J. Coal Prep. Util. **30**, 283–294 (2010). https://doi.org/10.1080/19392691003776814
74. Kademli, M., Gulsoy, O.Y.: The moisture content effect on coal cleaning performance of dry separator in different feed rates. Int. J. Coal Prep. Util. **41**, 463–473 (2019). https://doi.org/10.1080/19392699.2019.1639679

75. Ghosh, T., Honaker, R.Q., Patil, D., Parekh, B.K.: Upgrading low-rank coal using a dry, density-based separator technology. Int. J. Coal Prep. Util. **34**, 198–209 (2014). https://doi.org/10.1080/19392699.2014.869934
76. Davaasuren, J., Kim, B.-G., Lee, J.-H., Davaatseren, G., Bazarragchaa, M.: Dry coal preparation of fine particles by KAT process. In: XVIII International Coal Preparation Congress: 28 June–01 July 2016 Saint-Petersburg, Russia, pp. 1171–1176 (2016). https://doi.org/10.1007/978-3-319-40943-6_184
77. Sampaio, C.H., Ambrós, W.M., Cazacliu, B., Moncunill, J.O., José, D.S., Miltzarek, G.L., de Brum, I.A.S., Petter, C.O., Fernandes, E.Z., Oliveira, L.F.S.: Destoning the Moatize Coal Seam, Mozambique, by Dry Jigging. Minerals **10**, 771 (2020). https://doi.org/10.3390/MIN10090771
78. Mohanta, S., Meikap, B.C.: Influence of medium particle size on the separation performance of an air dense medium fluidized bed separator for coal cleaning. J. S. Afr. Inst. Min. Metall. **115**, 761–766 (2015). https://doi.org/10.17159/2411-9717/2015/V115N8A13
79. He, J., Bai, Q., Du, T.: Beneficiation and upgrading of coarse sized low-grade bauxite using a dry-based fluidized bed separator. Adv. Powder Technol. **31**, 181–189 (2020). https://doi.org/10.1016/J.APT.2019.10.009
80. Firdaus, M., O'Shea, J.-P., Oshitani, J., Franks, G.V.: Beneficiation of coarse coal ore in an air-fluidized bed dry dense-medium separator. Int. J. Coal Prep. Util. **32**, 276–289 (2012). https://doi.org/10.1080/19392699.2012.716801
81. He, J., Liu, C., Xie, J., Hong, P., Yao, Y.: Beneficiation of coarse particulate iron ore by using a dry density-based fluidized bed separator. Powder Technol. **319**, 346–355 (2017). https://doi.org/10.1016/J.POWTEC.2017.07.007
82. Franks, G.V., Firdaus, M., Oshitani, J.: Copper ore density separations by float/sink in a dry sand fluidised bed dense medium. Int. J. Miner. Process. **121**, 12–20 (2013). https://doi.org/10.1016/J.MINPRO.2013.02.008

Chapter 4
Electrostatic Separation

4.1 Introduction

The electrostatic separation technique is generally utilized for recycling metallic materials from electronic waste [1–6]. This dry separation technology is also successfully employed in mineral processing [7–10] and the food industry [11, 12]. As a perfectly dry method, electrostatic separation is an environmentally friendly technique that has received special attention due to the increasing scarcity of water resources. However, the capacity of this separation method for fine materials (<75 μm) is low. For effective operation, the input feed to most electric-based separation machines should be in a layer, decreasing the throughput as the small mineral particle size. This is the main reason for the capacity limitation when electrostatic separation is applied [13]. This technology fundamentally uses the differences between inherent electrical properties of various materials in a feed to separate them by utilizing a high electric voltage. According to the electrical conductivity, minerals can be categorized into three main groups: non-conductive (insulator), semi-conductive, and conductive. The electrical resistance strength would decrease from first class to the last one. Table 4.1 shows typical conductive and non-conductive minerals [14, 15].

For the electrical separation, two major forces can be considered; electrophoretic and dielectrophoretic forces. The former is the force that charged materials experience in an electric field, and the latter is occurred by neutral materials when exposed to a non-uniform electric field. However, using dielectrophoresis is practically uncommon in mineral processing. In mineral processing, prior to employing electrophoretic force for separation purposes, a pre-preparation stage is required to improve the process selectively. This step is performed based on differences in minerals' conductivity properties. The electrostatic separation method is generally considered a surface-based technique like froth flotation because electrical conduction mainly occurs in the atoms' surface layers [13, 14].

There are several mechanisms by which different materials can be electrically charged, including ion bombardment (corona charging), ion and electron-beam

S. C. Chelgani and A. Asimi Neisiani, *Dry Mineral Processing*,
https://doi.org/10.1007/978-3-030-93750-8_4

Table 4.1 Typical conductive and non-conductive minerals [14]

Non-conductive minerals	Conductive minerals
Apatite	Cassiterite
Barite	Chromite
Calcite	Diamond
Coal	Feldspar
Corundum	Galena
Garnet	Gold
Gypsum	Hematite
Kyanite	Ilmenite
Monazite	Limonite
Quartz	Magnetite
Scheelite	Pyrite
Sillimanite	Rutile
Spinel	Sphalerite
Tourmaline	Stibnite
Zircon	Wolframite

thermionics, freezing, conductive-induction, radioactive decay charging, triboelectric (contact electrification), photoelectric field emission, and photoelectric charging. However, the electric charging of mineral particles can be performed through three main mechanisms. Based on these mechanisms, three main types of separators have been developed:

- Ion bombardment (corona charging) mechanism
- Conductive-induction mechanism
- Frictional charging (tribo-charging or contact electrification) mechanism [10, 14, 16].

The choice of separator type should be proportional to the particle characteristics in the feed. For instance, when input feed comprises coal (as an almost non-conductive mineral), along with other minerals (conductive or insulator), choosing the separating method wholly depends on the inherent electrical features of coal and its associated impurities. While the triboelectric separating method can be utilized to treat insulator materials, the ion bombardment mechanism can recycle non-conductive materials from conductive ones. For both conductive-conductive and insulator-conductive separation, the conductive-induction mechanism can be successfully applied. However, only the triboelectric separators have been applied in the case of anthracite [16–18].

4.2 Ion Bombardment (Corona Charging) Mechanism

Corona charging by using ion bombardment is a mechanism in which a high electric field is applied between two electrodes to ionize the gas close to the electrodes and create corona discharge (a continuous flow of ions in the gaseous state). The gaseous ions bombard the mineral particles, causing an electric charge in the particle's surface and consequently separation. From the perspective of mineral processing, the difference in the inherent conductivity of various minerals leads to the different amount of electrical charge attained; therefore, mineral particles experience different forces. This mechanism is analogous and can be used to eliminate fine dust particles from gas streams in electrostatic precipitators. It has been demonstrated that the ion bombardment technique can effectively separate various materials [17, 18].

A typical corona charging-based separator is the high-tension roll (HTR), consisting of a rotating drum made from conducting material such as steel, corona discharge, AC corona electrodes, and a brush scraper (Fig. 4.1). HTR separators were initially utilized in the processing of sand containing heavy minerals [13]. It

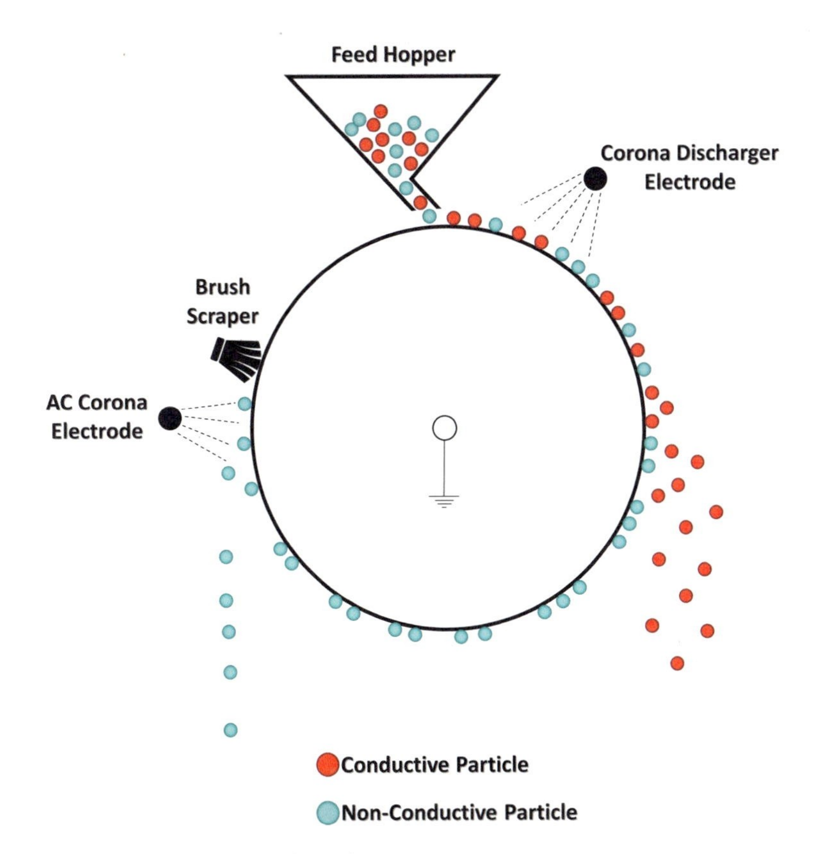

Fig. 4.1 Corona charging mechanism [19]

was then used to clean coal [19] and the separation of metals from plastic wastes [20]. As shown in Fig. 4.2, immediately after feeding to the rotating drum and charging the particles, conductive minerals discharge the ions that are received. In contrast, non-conductive particles retain the attained ions on their surface and pin to the drum (Fig. 4.2). Subsequently, centrifugal force, which is developed by rotational motion, separates conductive minerals from the feed. Insulator mineral particles are transported and detached eventually by AC corona electrode and brush scraper (Fig. 4.3) [17, 18, 21]. Due to the possibility of particles slipping on the drum surface, they

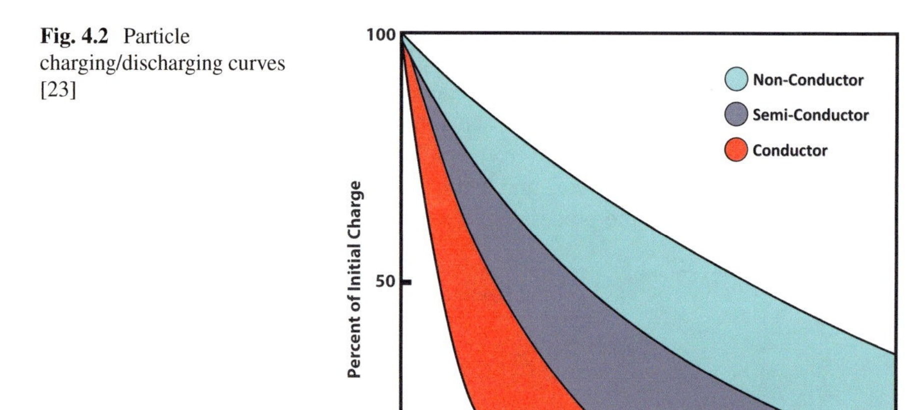

Fig. 4.2 Particle charging/discharging curves [23]

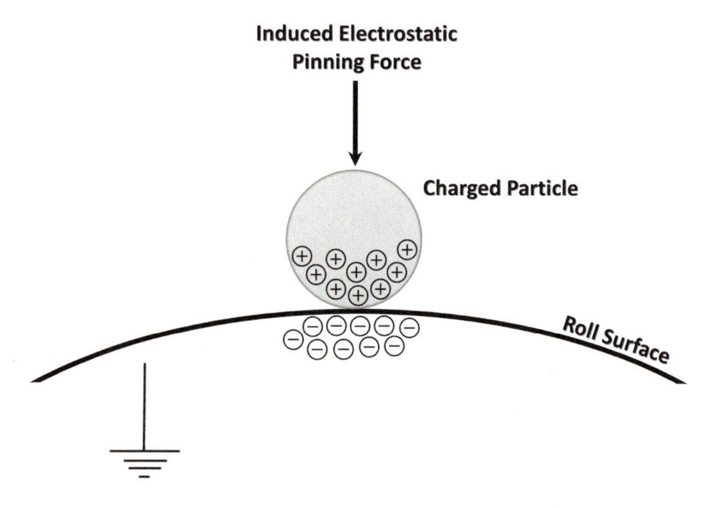

Fig. 4.3 Pinning force experienced by a non-conductive particle in HTR separators [21]

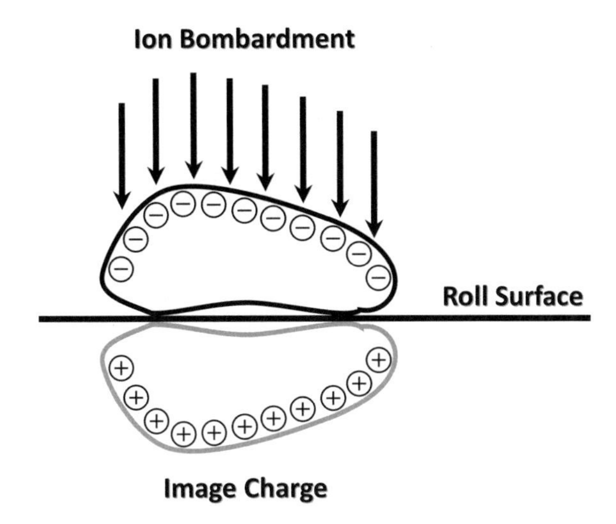

Fig. 4.4 The concept of image force for A non-conductive particle [20]

do not instantly accelerate to the drum speed. Therefore, it is difficult to predict the exact path of the particles [22].

High-tension separators can be utilized for input feeds with particle sizes in the range of 60–500 μm. Particle size considerably influences separation efficiency. For the large particle, less electrical charge can be attained on the surface relative to particle mass. Therefore, coarser non-conductive mineral particles are far more easily separated from the drum surface. On the other hand, finer particles receive more electrical charge. This causes, after the separation operation, some fine conductive particles to be present in the non-conductive fraction and vice versa. This cross-contamination has a negative effect on separation efficiency. This phenomenon can be interpreted by the interaction between the centrifugal force and the image force (Fig. 4.4), which attaches electrically charged mineral particles to the roll surface. The centrifugal force is a function of the particle mass, whereas the image force is influenced by changing particles' surface area [14].

Consequently, centrifugal force is dominant in terms of the coarse particle sizes [24, 25]. Alongside the particle size of the input feed, other parameters also affect the separation efficiency of electrostatic machines, including electrodes geometry, rotor diameter, rotational speed, voltage, electrode polarity, particle density, and splitter position [14]. The larger the rotor diameter, the more recovery will be achieved, whereas a smaller rotor diameter enhances the grade of the fraction of the conductive particles [24].

The effect of the rotor's rotational speed on the separation process can be studied from two perspectives. On the one hand, when the speed increases, mineral particles have more time to lose an electrical charge, increasing the probability of conductive particles in the non-conducting fraction. On the other hand, the higher the rotational speed, the greater the centrifugal force. That is why non-conductive mineral particles are more likely to be present in the conductive fraction when the rotational speed of the drum increases. As a result of this reciprocal effect, it was reported that the effect

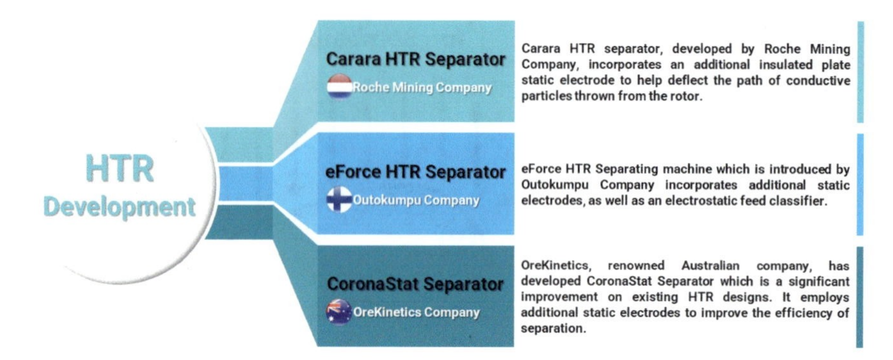

Fig. 4.5 Recently developed HTR machines [26]

of rotor speed on the separation efficiency is complicated and strongly dependent on the particle conductivity [24].

At least one mineral with strong inherent conductivity in the input feed is crucial when HTR is considered for the separation. It has been demonstrated that broad differences in mineral conductivities in a particular feed would not necessarily lead to an efficient separation. If all mineral species are non-conductive or weak-conductive, adverse effects on separation efficiency can be observed. On the contrary, two strongly conductive materials could be effectively separated with a slight difference in their inherent conductivities [2]. HTR separators have been developed in recent years, resulting in three new types of separators, which can be seen in Fig. 4.5 [26].

CoronaStat separator, developed by OreKinetics, is one of the latest versions of HTR electrostatic machines. This sophisticated device is equipped with additional static electrodes to enhance separation efficiency. Unlike mentioned separators, these electrodes are not exposed, leading to making the device far more secure to operate. Compared to traditional HTR machines, the most significant change made in the CoronaStat separator is the existence of induction electrodes, which enhance the attaching force on non-conducting mineral particles and accelerate the discharge rate for conductive particles simultaneously. Figure 4.6 demonstrates the simplified diagram of the CoronaStat separator [27, 28].

4.3 Conductive Induction Mechanism

Another mechanism utilized by electrostatic separators is conductive induction, in which an electrical field is developed in the space between a charged surface and a grounded plate. When mineral particles are transferred on the charged inclined surface, the polarization of materials occurs due to the existence of an electric field. Similar to what happens in HTR separators during electrical discharge, the most important factor in the separation process is the particle's electrical conductivity,

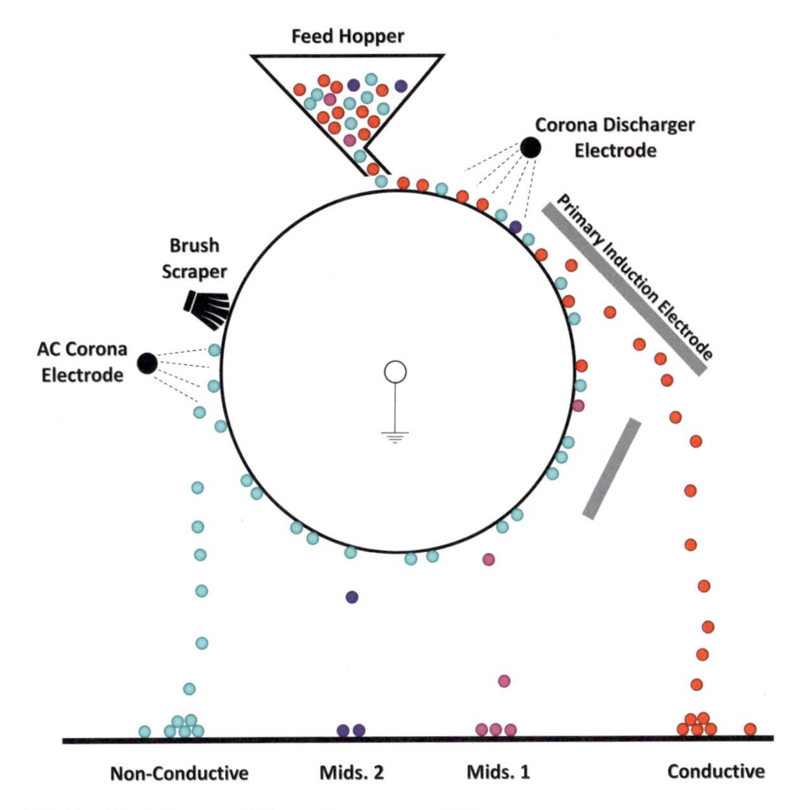

Fig. 4.6 Simplified diagram of CoronaSat separator [28]

which determines the capability of gaining voltage of charged plate and correspondingly the polarization intensity. After feeding and contacting mineral particles to the charged plate, polarization quickly occurs in the conductive materials [16, 29].

In this process, particles' amount of acquired electrical potential is equal to the charged surface potential. The insulator materials also obtain polarity; however, only that side of them attain surface polarity, which does not meet the charged surface. The other side of the non-conductive mineral that is in contact with the charged plate approximately acquires the same amount of positive and negative electrical charge; therefore, it does not possess polarity (Fig. 4.7). This phenomenon is the high electrical resistance of non-conductive particles, which results in a delay in charge distribution throughout the surface of particles. In contrast, insulator particles cannot redistribute obtained charges and consequently would remain polarized. In addition to the inherent conductivity of particles, the size and shape of the particles affect the polarization intensity and the resultant electric force [16, 23, 29, 30].

If a polarized non-conductive mineral particle, which possesses no net electrical charge, contacts a conductive surface, it does not experience attraction force from an

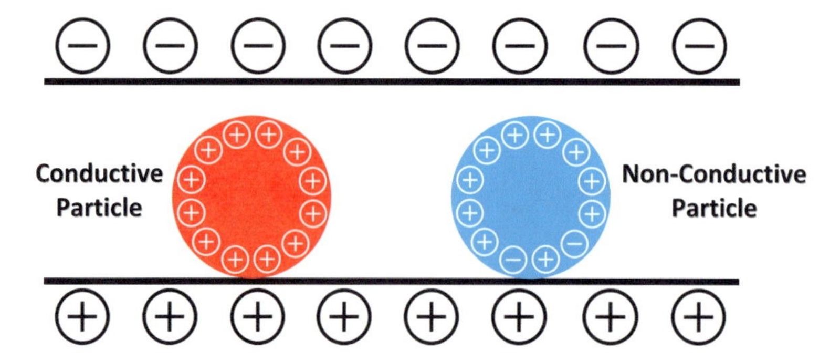

Fig. 4.7 The behaviour of conductive and non-conductive particles during conductive induction charging [29]

applied electric field. However, conductive minerals are attracted towards an electrode that provides the opposite electric charge [30]. As shown in Fig. 4.8, the conductive induction mechanism can be considered an electrostatic separation process that electrical charges are induced on uncharged conductive mineral particles by applying no attraction force on insulators with no net charge [14].

Machines that use the conductive induction separation mechanism are typically applied to separate a mixture of weak-conductive and strong-conductive minerals. For

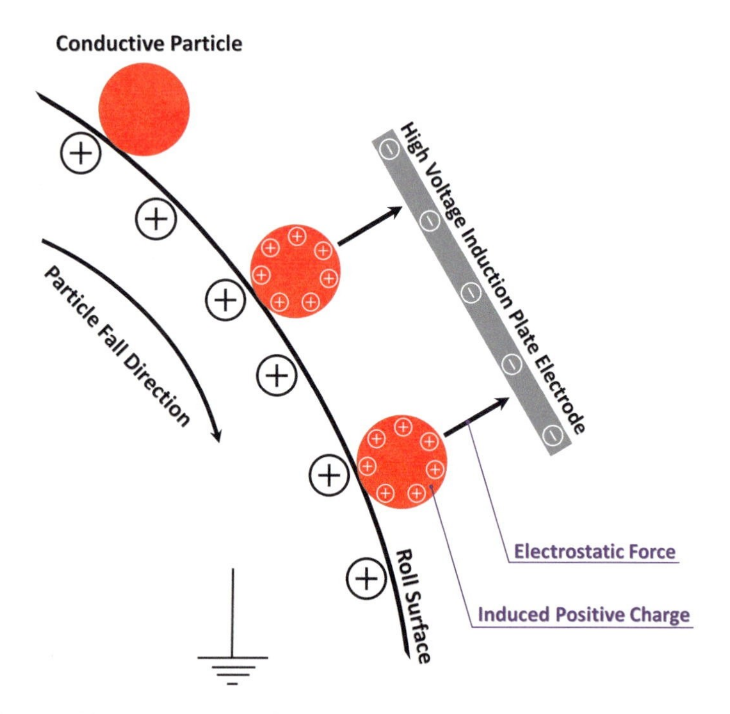

Fig. 4.8 A graphical representation of conductive induction mechanism [14]

this purpose, static electrodes are employed to lift electrically charged conductive minerals from the stream, whereas other minerals remain attached to the inclined surface. The electrostatic plate (ESP) separator is the most well-known separator using this mechanism. This separator is extensively used in sulfides roasting units to separate exhausting roasted dust present in the air stream. The humidity of the feed entering both ESP and HTR separators must always be controlled and kept low since extra moisture can change the conductivity of the particle surface and air (the fluid medium of the separator). The change in the particle surface conductivity in the presence of water can be due to either the dissolution of ions or the interaction of water molecules with each other. OreKinetics company, designer and developer of CoronaStat for HTR separation, has also introduced an enhanced ESP separator called UltraStat (Fig. 4.9). The main changes in the UltraStat are the geometry of the improved electrodes and the different feed stream paths. It also has equipped with extra electrodes and roll for the further enhancement in the lifting force and the latter to clean the surface of the main roll [13, 23].

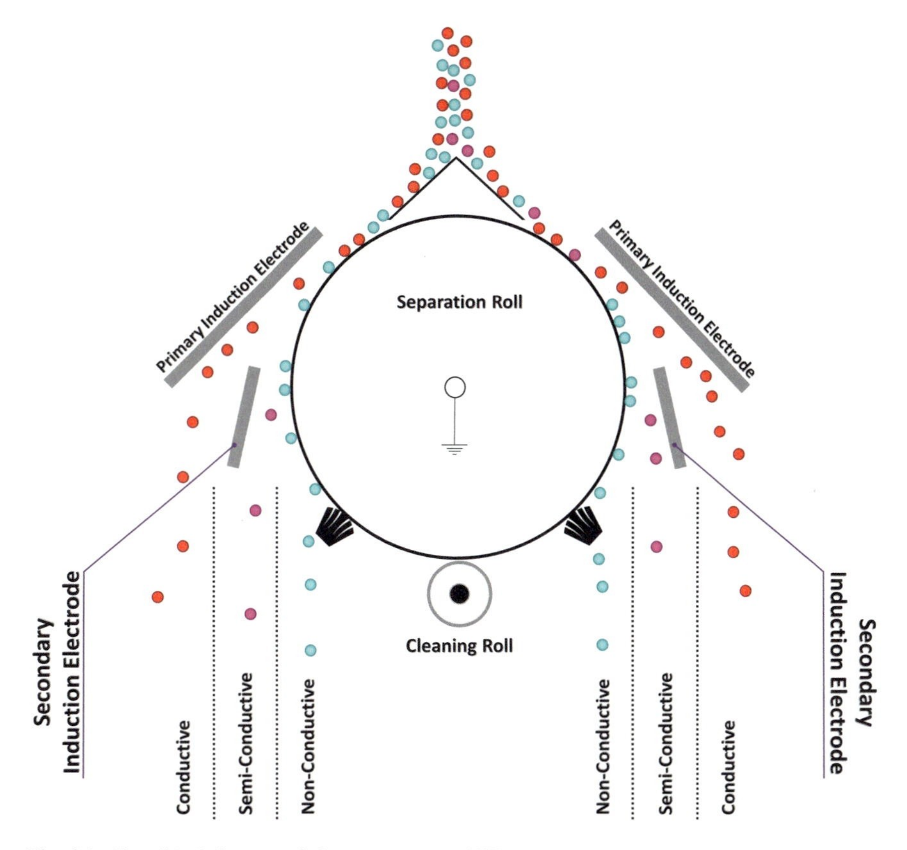

Fig. 4.9 Simplified diagram of ultrasat separator [13]

4.4 Triboelectric Charging Mechanism

The tribo-electrostatic charging mechanism is used in various areas, such as polymer recycling, powder coating, drugs delivery, electrophotography, fire hazard and safety, and the food industry. In mineral processing, this method has been successfully utilized for coal beneficiation [31–35] and the separation of various mixtures such as quartz- calcite [36] and quartz-feldspar [37]. Figure 4.10 shows the differences between triboelectric charging and other mechanisms [29].

The differences in the surface work function of various non-conductive materials are the basis of the triboelectric (contact electrification) charging mechanism. This mechanism leads to collision and friction between particles, polarity, and power variety [38]. Work function is the minimum amount of energy required to detach an electron from the Fermi level and transfer it to a vacuum level. Additionally, Fermi level is defined as the energy level in which half of the energy states in a material are occupied by electrons. A particle with a low Fermi level possesses a greater surface work function than a material with a higher Fermi level. During the contact of two different particles, the one with a smaller surface work function donates electrons; therefore, its electrical charge becomes positive. In contrast, the particle with a higher surface work function obtains the electrons and becomes negatively charged. After completing this process, the level of two species of minerals equalizes. It should also be noted that the electrical conductivity of most minerals is in the range of 10^{-8} and 10^{5} Ωm^{-1}, and they can be categorized as semi-conductors. However, the triboelectric separation method can be immensely applied while a broad difference in conductivity is not required in this technique. This method is applicable in the case of almost every mixture that its particles have a difference in Fermi level [14, 39]. As shown in Fig. 4.11, the triboelectric separator comprises a horizontal section called tribo-charger, where there is a continuous flow of air and mineral particles. This section, also known as the friction-charging section, mineral particles are polarized after either contacting embedded plates or rubbing to each other. Subsequently, particles pass through a nozzle and enter the high voltage processing section (free-fall chamber), in which there are positive and negative plates creating an electrical field. The path of mineral particles inside the processing zone depends on two main factors:

- the charge of particle surface (negative or positive)
- charge-mass ratio (surface charge per particle mass unit).

The negative plate attracts the positively charged minerals particles, whereas the negative charge particles are deflected towards the positive plate. During this step, minerals particles that possess a higher charge-mass ratio experience more electrical force and move faster. Notably, the particles with a too low charge-mass ratio follow neither negatively charged particles nor the other one stream, causing a decrease in separation efficiency [33, 38].

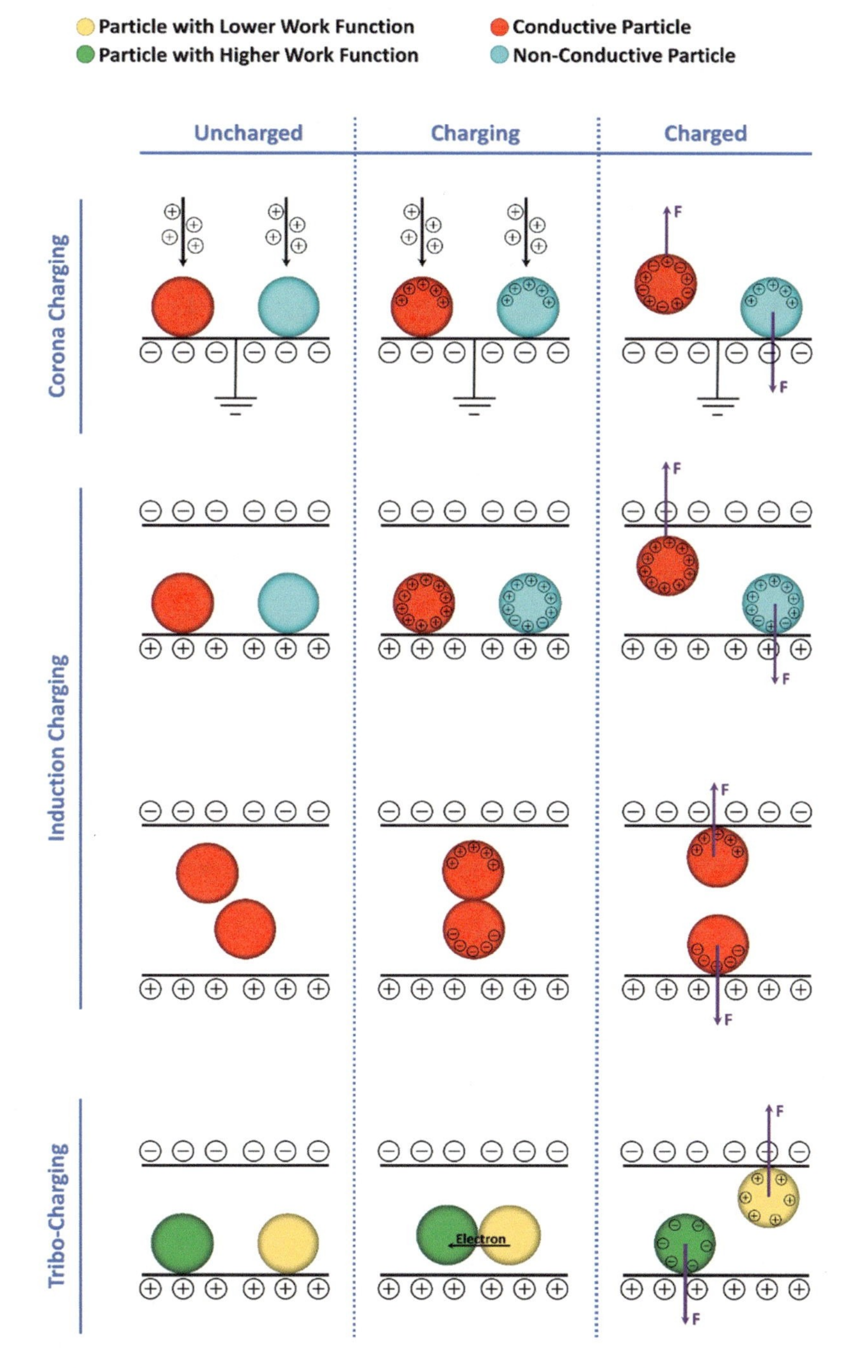

Fig. 4.10 Comparison of different electrostatic charging mechanisms [29]

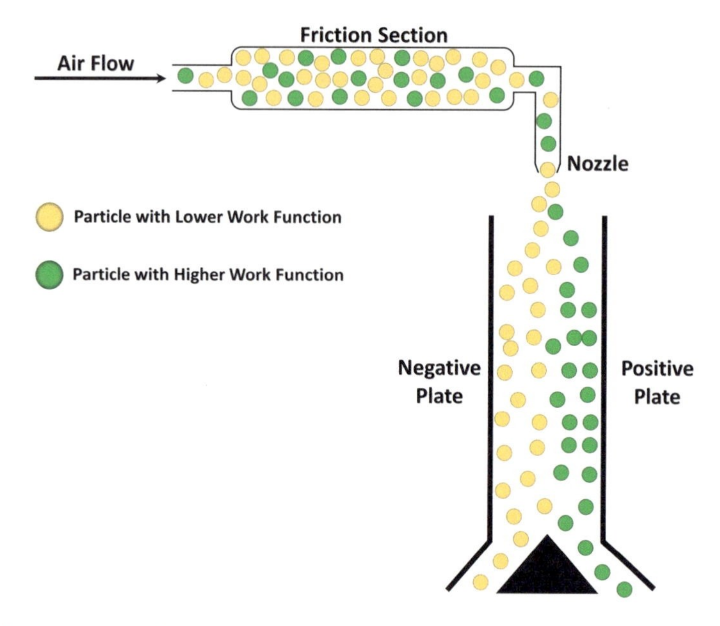

Fig. 4.11 Triboelectric charging separation mechanism [38]

One of the most common uses of triboelectric separation is in coal beneficiation. Increasingly strict environmental laws have forced industries to use high-grade (low-ash) coals. The main reason is that the use of low-grade coal increases the emission of SO_X and NO_X. Coal cleaning using the tribo-electrostatic separation method can offer environmental and commercial benefits to coal beneficiation units and coal users because this method can effectively improve the quality alongside the commercial value of coal [32, 34, 40].

In order to make an effective separation with the help of a triboelectric separator, the affecting parameters are as follows:

- *Surface Treatment*: Before using tribo-charging, raw materials are sometimes pre-prepared in order to control and improve the charge attainment. These processes include chemical pretreatment, surface cleaning, irradiation, thermal pretreatment, and surface doping, enhancing the electrical charge differences in various minerals.
- *Air Flow Rate*: In both charging and separation steps, airflow can play a key role during charging. Increasing the airflow until its optimal level improves the yield percentage of the final product and proportionately reduces the ash content. The main reason for the yield improvement is that the particle's energy is utterly dependent on the airflow rate in the friction-charging section. The more the intensity of the airflow leads to the higher the particle velocity. Therefore, particle-plate or particle-particle friction will be improved and provide a more charge-mass ratio. However, if this airflow rate exceeds the optimal point, the ash content increase,

and yield decreases. The airflow regime plays an important role in the friction-charging process, and if a turbulent regime is generated, the process will not continue properly. Accordingly, the optimum airflow rate is the speed at which contacting energy is at the highest level, and the turbulence is minimum.

- *Air Flow Humidity*: The presence of moisture in the airflow causes the surface of the mineral particles to become wet and consequently increase their conductivity. Hence, the ability of mineral particles to retain surface electrical charge considerably decreases, which leads to a decrease in charge-mass ratio. This issue can cause major difficulties for the separation process and reduce yield percentage.
- *Particle Size*: Mineral particle size is another affecting variable in triboelectric separators is particle size. In the case of a specific mineral, the smaller the particle size, the higher the surface work function. Since differences in work function are the basis of this method, particle size plays an important role in separation efficiency.
- *Separation Voltage*: The increase in electrical potential enhances the attraction force intensity. Therefore, embedded plates in the chamber can capture more particles, leading to improving yield percentage.
- *The length of Separation Path*: The movement velocity of mineral particles towards the oppositely charged plate is considerably dependent on the amount of charge-mass ratio. Additionally, suppose the separation path length is too short. In that case, mineral particles, more specifically those with a low charge-mass ratio, do not have enough time to be captured by the electrode. This may result in the presence of impurity particles in the product path. On the contrary, an excessive increase in separation zone length would increase particles' residence time, leading to turbulent regime generation, and consequently, mineral particles deviate from their favorable path.
- *The Position of Splitter*: The last part of the triboelectric separation machine is the splitter, located at the bottom of the processing chamber. In this part, the streams of various mineral particles are separated and collected as products or tailings according to their electrical charge. If the splitter is shifted towards the tailing collection section, more particles will be collected in the final product part, which increases the yield and ash content and vice versa.
- *Gangue Minerals*: A mineral particle can be either positively or negatively charged depending on other minerals present in the mixture. This is because the work functions of gangue minerals can be higher or lower than valuable minerals. For example, during a beneficiation investigation, ilmenite acquired different electrical charges through the triboelectric charging method in different experiments when gangue minerals varied.
- *Flowsheet*: Similar to other mineral processing methods, the triboelectric technique can be utilized as a single-stage or multi-stage procedure. For instance, materials are initially fed to the separator during coal beneficiation and separated into first-stage products and tailings. After that, this product is reprocessed with another triboelectric separator, and a much cleaner product would be

obtained, containing fewer impurities than the initial product. However, a proportion of middling particles-impurities are extracted during the secondary process [37, 39, 41–45].

Triboelectric separation, which has received more attention nowadays, has been improved over time, and various types of separators have been developed based on this mechanism:

4.4.1 Double Drum Separator

The separation principle in the double drum separator is based on the surface electrical charge of different minerals, similar to previously described machines. However, the charging process differs slightly from what occurs in conventional triboelectric machines. In this type of device, particles face a rotating brush rather than the plates, resulting in more agitation. Therefore, the probability of particle-particle contacts and collisions inside the charging chamber increases. According to the surface work function, particles are negatively or positively charged. Because there is no plate in the charging chamber, the charging procedure is only carried out through particle-particle contacts. Subsequently, the electrically charged particles are transferred by airflow towards the separation section, which comprises two insulator drums. Two electrodes are embedded on two different sides of each drum and possess opposite electrical charge signs, separated by a non-conductive spacer [46, 47].

Thus, one of the electrodes (capturing electrode) directs the mineral particles towards the drum, whereas the second electrode (detaching electrode) moves the particles away from the drum surface by applying repulsive force. During this process, after entering the charged particles into the electrical field, the electrode with the opposite electrical charge captures and attaches it to the drum, which is rotating in the clockwise direction. After the attached particle reaches the second electrode, a brush next to it, the particle is detached from the roll surface by applying repulsion force in conjunction with a brush and eventually entering the corresponding box. Middling mineral particles usually acquire a low electrical charge; therefore, they are less likely to be attracted. As seen in Fig. 4.12, these mineral species are removed by a part called "frit" embedded at the end of the separation chamber. During an experiment, a mixture of anthracite and quartz was treated using a double drum triboelectric separator. By changing several variables such as electric field strength and particle size, separator efficiency was evaluated. Since quartz possesses a higher surface work function in comparison to anthracite, quartz and anthracite always acquire negative and positive surface charges, respectively. If electric field strength and particle size increase, the coal (anthracite) recovery improves, but quartz particles are less attracted by the electrode [46, 47].

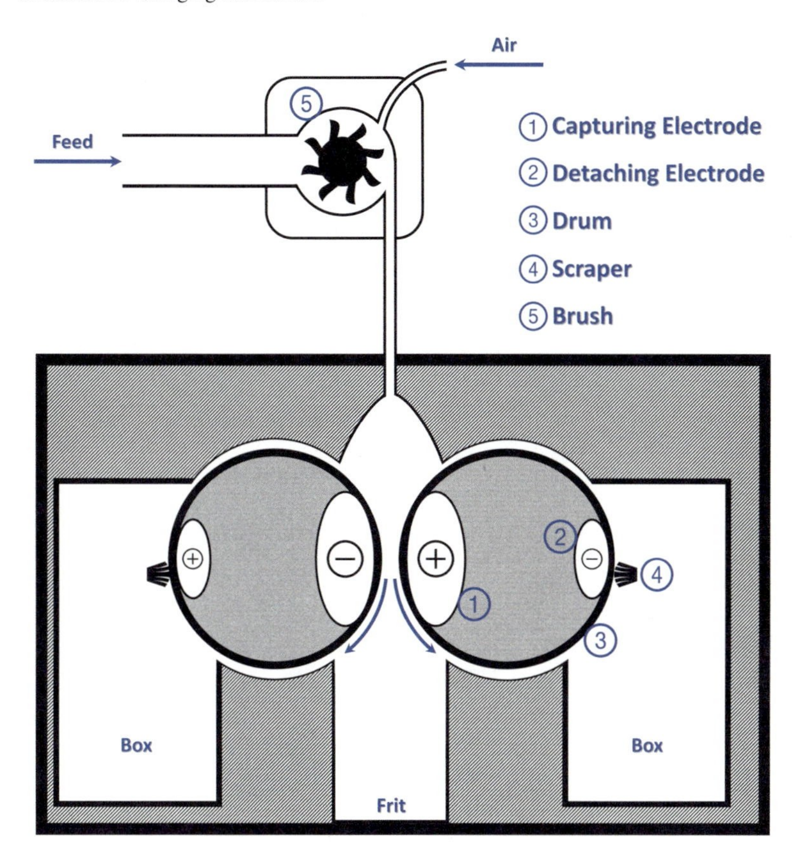

Fig. 4.12 Double drum separator [46]

4.4.2 Rotary Tribo-Electrostatic Separator

Rotary Tribo-electrostatic Separator (RTS) is an updated version of the triboelectric separator, which the University of Kentucky has been developed. Design improvements in this machine have made the particle charging process more efficient. The most important improvement is a high voltage rotating plate revolving inside a ground-connected cylindrical, which is an embedded chamber instead of a fixed charging plate. In order to avoid generating a turbulent regime inside the separation zone, the airflow rate has been significantly diminished. In this upgraded RTS, the probability of particles colliding with each other is much higher than the conventional types.

Furthermore, the revolving charger rotates at high speed, leading to more friction energy experienced by particles. The rotary charged plate significantly enhances the charge exchange between various minerals and plates. In this machine, the charge density of the mineral's surface has been considerably improved compared to the previously mentioned charging mechanism. Hence, the amount of weakly charged

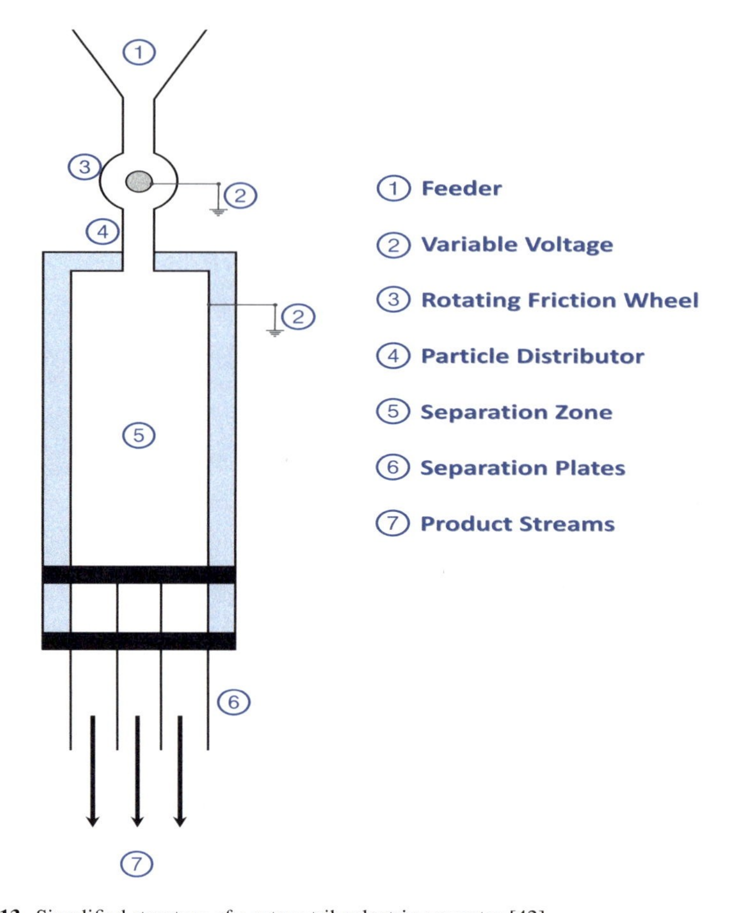

Fig. 4.13 Simplified structure of a rotary triboelectric separator [42]

particles (middling mineral particles) would be greatly decreased [17, 48, 49]. As an efficient and satisfactory dry separation technique, this method has been highly rewarded in the coal beneficiation units [8, 50, 51]. Figures 4.13 and 4.14 demonstrate a rotary triboelectric separator's structure and its working principle, respectively [42].

4.4.3 Triboelectric Belt Separator

A triboelectric belt separator, also known as a ST separator, has been recently introduced to the processing units. During separation by ST separator, raw materials are fed into a narrow gap (<1.5 cm) between two electrodes with opposite charges. Interparticle contact causes materials to charge triboelectrically. In the subsequent step, charged particles are separated by an open-mesh belt which travels in a continuous loop at a speed of 5–20 m s^{-1}. The moving belt transfers the mineral particles next

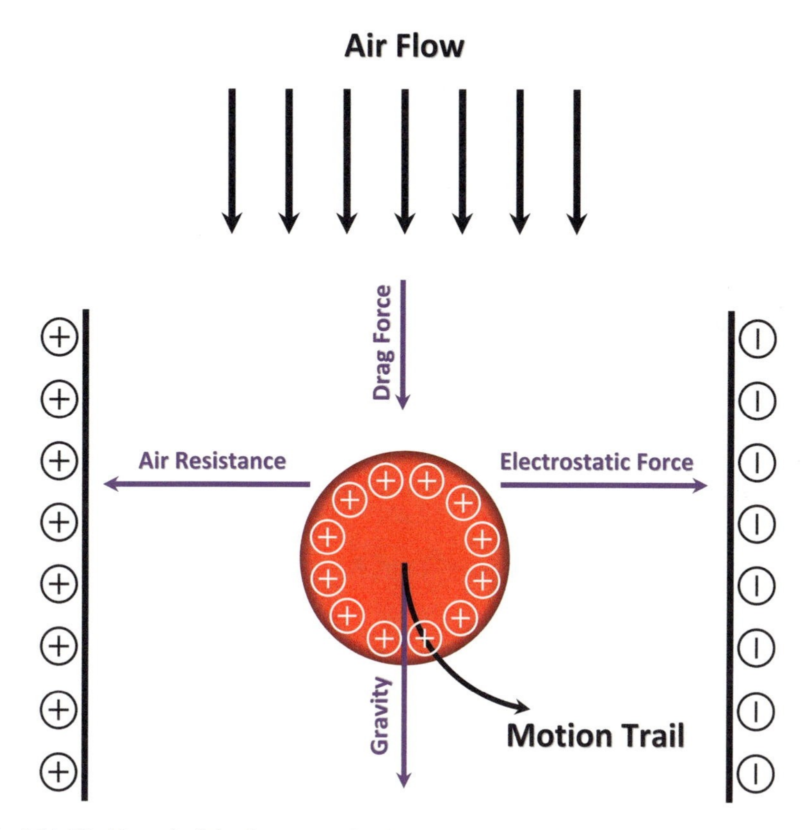

Fig. 4.14 Working principle of a rotary triboelectric separator [42]

to each electrode towards opposite ends of the machine. The counter-current stream, its highly turbulent pattern, and continual charging by a triboelectric mechanism can lead to an efficient separation with satisfactory recovery and grade even in a single-stage separation. Figure 4.15 shows a schematic diagram of the ST triboelectric separator. Despite its comparatively simple design, the triboelectric belt separator has significant advantages:

- The process is completely dry.
- It requires no additional materials.
- It produces no wastewater or air emissions.
- Its compactness permits flexibility in installation designs.
- It can separate fine particles at the range of 1–300 μm [36].

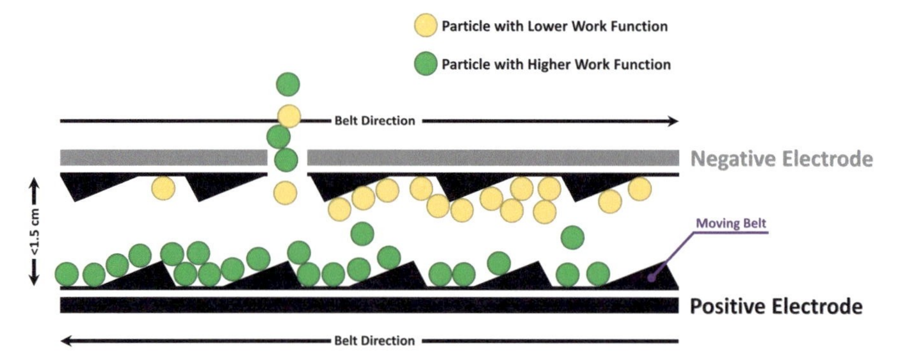

Fig. 4.15 A schematic diagram of ST triboelectric separator [36]

4.5 The Applications of Electrostatic Separation

Electrostatic methods are used to separate various mineral and non-mineral mixtures. Some of them are mentioned below:

4.5.1 Coal Beneficiation

Coal is the primary energy source in numerous countries, and its important role is expected to remain unchanged for years. Approximately 80% of coal is burned to produce energy, which causes a significant amount of sulfur dioxide and smoke dust to be produced in the dust. For this reason, it is very important to produce high-quality coal that has a small amount of impurity. Severe environmental problems have occurred in many countries due to the burning of unprocessed coal, and the main reason for those disasters is water shortage. For instance, in China, which is the largest producer and consumer of coal globally, more than 65% of coal reserves are located in areas facing water shortages [35, 52].

Dry coal beneficiation techniques, such as handpicking, magnetic separation, frictional separation, electric separation, microwave separation, air-dense medium fluidized bed beneficiation, etc., are applied to reduce or eliminate water requirements in coal beneficiation. In recent years, electrostatic-based separation methods have been considered, and among the mentioned mechanisms, tribo-charging methods have received more attention. Flexibility in design and installation, no need for water, the ability to separate fine particles, and satisfactory efficiency and recovery are the most important advantages of using this method. For example, it is reported that for decarbonization of coal-series kaolin, using triboelectric separation can significantly enhance the recovery with acceptable removal of impurities, such as TiO_2 and Fe_2O_3 [35, 52].

4.5.2 Waste Printed Circuit Boards (WPCB)

As a result of technology and economic development, the amount of waste electrical and electronic equipment (WEEE) has been increasing rapidly at an alarming rate. Waste printed circuit board, composed of ceramic, metal, polymer, and other substances, makes around 6% of WEEE mass. WPCBs inappropriate treatment can cause environmental problems such as soil and water pollution. Additionally, WPCB possesses a considerable amount of valuable metals that can be recycled through mechanical, pyrometallurgy, hydrometallurgy, and biotechnology processes. Electrostatic separation is an effective and environmentally friendly method to separate different components of WPCB. For this purpose, one of the most effective machines is a two-roll separator developed based on the ion bombardment (corona charging) mechanism. Figure 4.16 shows a schematic representation of the two-roll type corona-electrostatic separator [53–56].

4.5.3 Sand Deposits

Electrostatic separation effectively separates various heavy minerals present in titaniferous sedimentary deposits and beach sands. HTR separators are used to process the sand deposits. In this process, several operating variables, such as feed rate, potential, the rotational speed of the roll, electrode geometry, and particle size distribution, should be concerned to achieve an acceptable separation. This procedure usually occurs in three steps. In the first stage, low-density minerals such as quartz and garnet, which make up around 90% of the feed, are rejected through wet gravity separation. Subsequently, wet magnetic separation eliminates magnetite and divides other minerals into magnetic and non-magnetic minerals. Finally, dry electrostatic separation is used to complete the process. As shown in Fig. 4.17, after drying magnetic and non-magnetic streams, they are separated by dry electrostatic separators. Therefore, each part is divided into conductive and non-conductive streams [14, 21, 22].

4.6 Main Applications and Producers

Table 4.2 summarizes the various application of Electrostatic Separation for different minerals. Table 4.3 shows different Electrostatic separator producers.

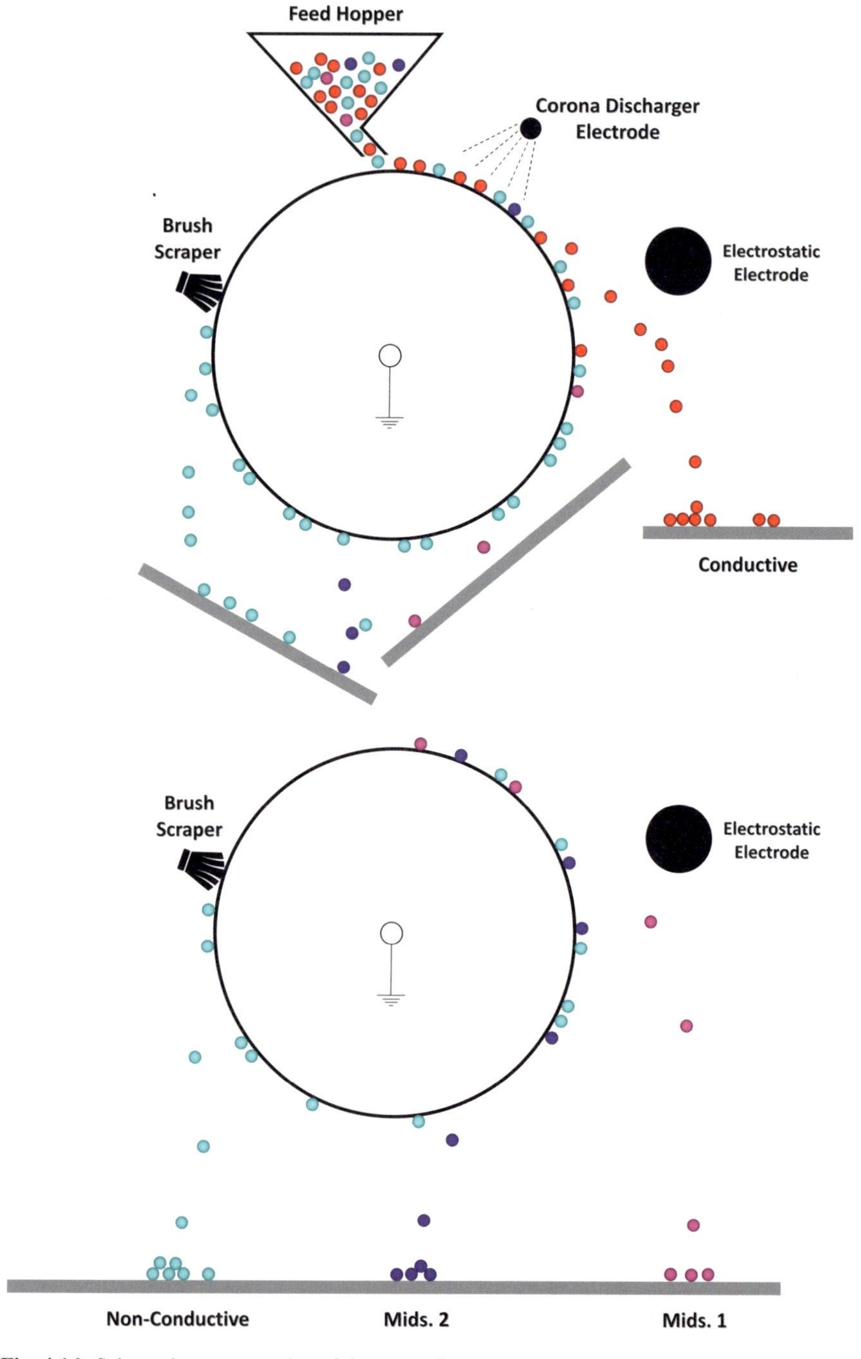

Fig. 4.16 Schematic representation of the two-roll type corona-electrostatic separator [3]

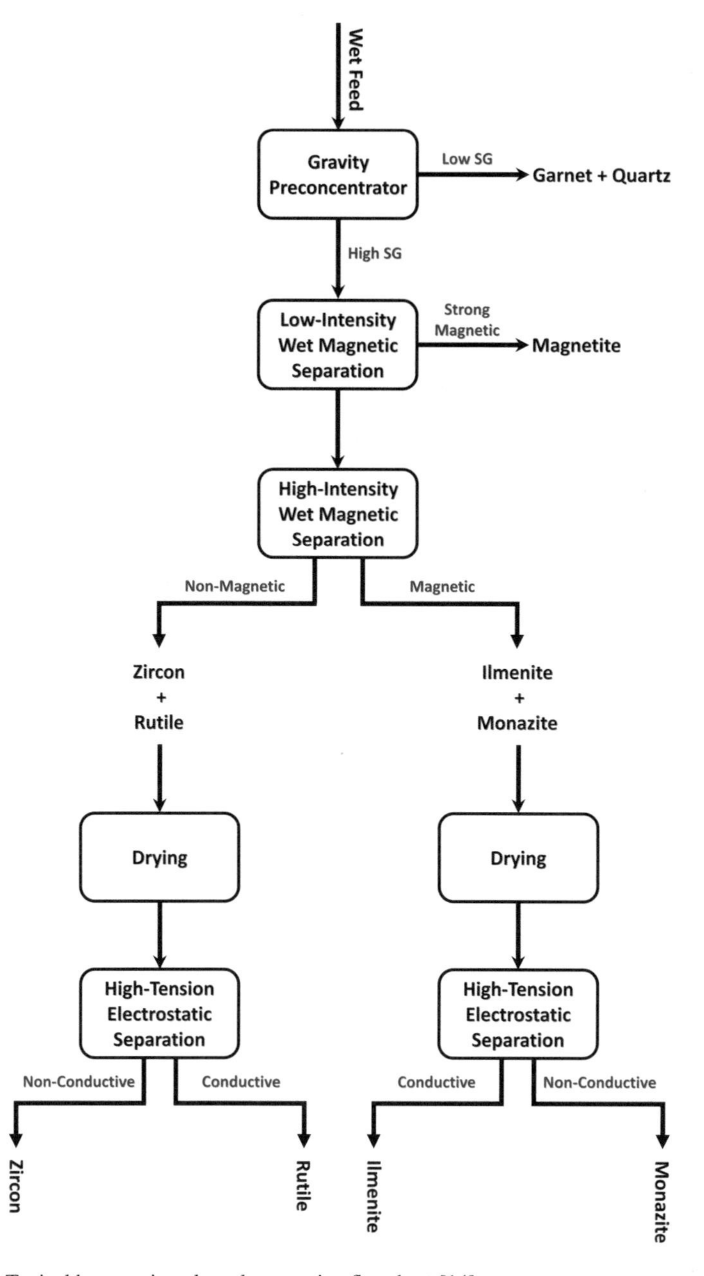

Fig. 4.17 Typical heavy mineral sand processing flowsheet [14]

Table 4.2 Summary of the investigations conducted on electrostatic for separation of various minerals

Separator	Feed	Parameters	Purpose	Results	References
Triboelectric separator	Waste Printed Circuit Boards	Feed size: 0.09–1.4 mm Feed rate: 300 gr min^{-1} Voltage: 30 kV Charger rotation: 200 rpm	Removal of inorganics contaminants from the nonmetallic fraction to improve their usability	– The content of organics increases from a negative product to a positive product, fluctuating from 48.90 to 75.66%	[57]
Drum Separator	Colemanite and Ulexite	Feed size: 0.5–1 mm Feed rate: 1000 gr min^{-1} Voltage: 10–22 kV Pre-heating: 0–65 °C Drum rotation: 60 rpm	Separation of boron minerals (colemanite and ulexite) from each other	– By changing the amount of voltage, the conductivity features of boron minerals vary – The amount of ulexite mineral particles attached to the separating drum was 92% with heating at 65 °C and at 22 kV applied voltage	[7]
Corona Discharge	Printed Circuit Boards	Feed size: 0.3–2 mm Drum rotation: 60 rpm Voltage: 30 kV	Metal content recovery	– Particle sizes between 0.6 and 1.2 mm are suitable for separation in industrial applications	[18]

(continued)

Table 4.2 (continued)

Separator	Feed	Parameters	Purpose	Results	References
Triboelectric separator	The mixture of various types of plastics	Feed size: 1.68–2.38 mm Electric field: 400 kV m^{-1} Ambient temp.: 30 °C Rubbing time: 60 s	Sorting plastics from a multi-component plastic mixture, namely acrylonitrile-butadiene-styrene (ABS), polypropylene (PP), and polyvinyl chloride (PVC)	– Products of ABS, PP and PVC, which can be achieved with a grade of 94.3%, 92.5% and 93.7%, respectively, are collected with a percentage mass distribution higher than 76.7% for them	[58]
Triboelectric separator	Coal	Feed size: −74 μm Feed rate: 454–4540 gr h^{-1} Gas velocity: 15 cm s^{-1} Voltage: 25 kV	Reduce ash contents from various Slovakian coal	- Maximum Ash recovery: 72% for Handlova Concentrate 70% for Ci'gel Concentrate 67% for Nova'ky Concentrate	[9]
Triboelectric separator	Plastic Wastes	Feed size: 0–6 mm Rotation: 300 rpm Voltage: 10–30 kV	Recycling of plastics (PET, PVC) from various wastes	– The optimum results were obtained at the potential greater than 20 kV and with a splitter position at 2 cm away from the center to the negative electrode. Under these conditions, the recovery of 98% and 98% is attainable for PET and PVC, respectively	[59]

(continued)

Table 4.2 (continued)

Separator	Feed	Parameters	Purpose	Results	References
Triboelectric separator	Kyanite Ores	Feed size: 0–0.315 mm Voltage: 35 kV	Evaluation of the advantages of replacing dry electrostatic method with flotation	– Concentrates with 59–60% Al_2O_3 (up to 95% for the mineral) and a pure quartz tailing can be obtained – It was established that unit energy costs for the dry technology are cut to 15–25 kWh/ton	[60]
Triboelectric separator	Coal–Silica Mixture	Feed size: 0.15–1 mm Rotation: 2000–5000 rpm Feed rate: 0.9–13.6 kg h^{-1} Injection air: 0.6–4.3 m s^{-1} Co-flow rate: 1.5–2.4 m s^{-1}	Separating Silica minerals from coals	– Optimum operating parameters were determined as: Feed rate: 9.1 kg t^{-1} Charger rotating speed: 4000 rpm Applied charger voltage: 5 kV Injection air rate: 0.6 m s^{-1} Co-flow rate: 2.1 m s^{-1}	[48]

(continued)

Table 4.2 (continued)

Separator	Feed	Parameters	Purpose	Results	References
Triboelectric separator	Pulverized Coal	Feed size: −74 μm Charger rotation: 4000 rpm Voltage: 40 kV Modifier dosage: 9 kg ton^{-1} Airflow velocity: 3.0 m s^{-1}	Exploring the effect of pulverized coal surface modification (by kerosene, diesel, acetic acid, and sodium stearate) on separation	– Kerosene has the best modification effect for the separation of pulverized coal – Using Kerosene deash rate, desulfurization rate, and recovery are increased by 1.87%, 0.44%, and 6.07%, respectively Moreover, ash content is reduced by 1.95%	[34]
Triboelectric separator	Coal	Feed size: 0–175 micron Charger rotation: 4310 rpm Voltage: 35 kV	Coal decarbonization	– The fraction size of 37–74 μm can be purified most effectively by the electrostatic separation method with the high yield and most significant reduction in LOI – For this fraction size, decarburization rate and decarburization efficiency index of 38.93% and 120.83% can be achieved	[61]

(continued)

Table 4.2 (continued)

Separator	Feed	Parameters	Purpose	Results	References
Two-roll corona separator	Printed Circuit Board Waste	Feed size: 0.09–1.2 mm Voltage: 20 and 30 kV	Separating multi-size granule of crushed printed circuit board waste	– Using a two-roll corona electrostatic separator, compared to roll-type, the conductive products increase by 8.9%, the middling products decrease by 45% The production capacity also increases by 50%	[62]
Corona separator	Waste Photovoltaic Cells	Feed size: 0.25–0.42 mm Voltage: 15–35 kV Drum rotation: 45–105 rpm Temperature: 21–26 °C	Recycling silver and silicon from waste photovoltaic cells	– The silver and silicon separation efficiencies increased with the voltage – The separation efficiency for silver and silicon was found to reach 96% and 98%, respectively	[3]

(continued)

Table 4.2 (continued)

Separator	Feed	Parameters	Purpose	Results	References
Corona separator	The residue of Electric Cables Recycling Process	Feed size: 1–2 mm Voltage: 25–26.5 kV	Separating aluminum from the residue of the electric cables recycling process	– Using a corona separator, it is possible to obtain a concentrate with the Al mass fraction of 97.86% at 91.20% recovery – The Splitter position has the most significant influence on the grade of concentrate and recovery The greatest recovery is achieved with the splitter position ranging from 60° to 100° Roll speed, individually, and the interaction of the roll speed and splitter position significantly influence the concentrate grade	[63]

(continued)

Table 4.2 (continued)

Separator	Feed	Parameters	Purpose	Results	References
Corona Separator	Coal	Feed size: 106–1000 μm Voltage: 10–20 kV Pre-heating: 60–100 °C Test atmosphere: – CO_2 – NO_2 – Compressed air	Removing sulfur, in the form of pyrite (FeS_2) from coal	In the optimum test, an ash reduction of 50.64% was achieved in a compressed air atmosphere using the 106–1000 μm size fraction particles – CO_2 and NO_2 gave results that compare favorably with the air atmosphere. Thus, they can be used for fine (−90 μm) coal beneficiation to reduce the explosion risk	[19]
Corona separator	Spent Lithium-Ion Batteries	Feed size: 0.212–5 mm Drum rotation: 10–50 rpm Voltage: 15–30 kV Deflector tilt: 0°–7.5°	Recovery of valuable materials from spent lithium-ion batteries	– The best-operating conditions for the electrostatic separator were a roll rotation speed of 20 rpm, electrode voltage of 25 kV, and deflector inclination angle of 0° – In the optimum test, the recovery of metals (conductive fraction) and polymers (non-conductive fraction) was 98.98% and 99.6%, respectively	[64]

(continued)

Table 4.2 (continued)

Separator	Feed	Parameters	Purpose	Results	References
Triboelectric separator	Ilmenite-Quartz Mixture	Feed rate: 2–12 g s^{-1} Voltage: 10–80 kV Airflow: 20–80 m^3 h^{-1}	Separating ilmenite from quartz	- A maximum ilmenite recovery of 51.71% with ilmenite content of 32.72% was obtained at 40 $m^3.h^{-1}$ airflow rate, 6 $g.s^{-1}$ feed rate, and 20 kV voltage	[65]
Triboelectric belt separator	$CaCO_3$-SiO_2 Mixture	$CaCO_3$: 90.5% SiO_2: 9.5%	Separating silica from calcium carbonate	– The recovery of 89% for $CaCO_3$ can be achieved and recovered product composition is 99.1% $CaCO_3$ and 0.9% SiO_2	[36]
Triboelectric belt separator	Talc-Magnesite Mixture	Talc: 58% Magnesite: 42%	Separating magnesite from Talc	– The recovery of 77% for Talc can be achieved and recovered product composition is 95% Talc and 5% Magnesite	[36]

Table 4.3 The main electrostatic separator producers

Equipment	Manufacturer	Location
Electrostatic separator	Allcontrols	UK
	Bunting	UK
	Das Env. Expert	Germany
	Dragoelectronica	Spain
	Erga	Russia
	Eriez	UK
	Hamos	UK
	Jinpeng	China
	Keller	USA
	Novomof	Switzerland
	Ore Kinetics	Australia
	Outokumpu	Finland
	Prodecologia	Ukraine
	R&R Beth	Germany
	Redoma	Sweden
	Roche	Netherlands
	Scheuch	Austria
	Spirotech	Netherlands
	Taizhou	China
	Wamatech	Denmark
	Zhengzhou	China

References

1. Franke, D., Suponik, T., Nuckowski, P.M., Golombek, K., Hyra, K.: Recovery of metals from printed circuit boards by means of electrostatic separation. Manag. Syst. Prod. Eng. **28**, 213–219 (2020). https://doi.org/10.2478/mspe-2020-0031
2. Hamerski, F., Krummenauer, A., Bernardes, A.M., Veit, H.M.: Improved settings of a corona-electrostatic separator for copper concentration from waste printed circuit boards. J. Environ. Chem. Eng. **7** (2019). https://doi.org/10.1016/j.jece.2019.102896
3. Zhang, Z., Sun, B., Yang, J., Wei, Y., He, S.: Electrostatic separation for recycling silver, silicon and polyethylene terephthalate from waste photovoltaic cells. Mod. Phys. Lett. B **31**, 1–11 (2017). https://doi.org/10.1142/S0217984917500877
4. Hamerski, F., Bernardes, D.P., Veit, H.M.: Operational conditions of an electrostatic separator for concentrate copper from electronic waste. Rev. Esc. Minas **71**, 431–436 (2018). https://doi.org/10.1590/0370-44672017710159
5. Lu, H., Li, J., Guo, J., Xu, Z.: Movement behavior in electrostatic separation: recycling of metal materials from waste printed circuit board. J. Mater. Process. Technol. **197**, 101–108 (2008). https://doi.org/10.1016/j.jmatprotec.2007.06.004
6. Wu, J., Li, J., Xu, Z.: Electrostatic separation for recovering metals and nonmetals from waste printed circuit board: problems and improvements. Environ. Sci. Technol. **42**, 5272–5276 (2008). https://doi.org/10.1021/es800868m

7. Eskibalci, M.F., Ozkan, S.G.: An investigation of effect of microwave energy on electrostatic separation of colemanite and ulexite. Miner. Eng. **31**, 90–97 (2012). https://doi.org/10.1016/j.mineng.2012.01.018
8. Bada, S.O., Tao, D., Honaker, R.Q., Falcon, L.M., Falcon, R.M.S.: A study of rotary tribo-electrostatic separation of South African fine coal. Int. J. Coal Prep. Util. **30**, 154–172 (2010). https://doi.org/10.1080/19392699.2010.497100
9. Soong, Y., Link, T.A., Schoffstall, M.R., Gray, M.L., Fauth, D.J., Knoer, J.P., Jones, J.R., Gamwo, I.K.: Dry beneficiation of Slovakian coal. Fuel Process. Technol. **72**, 185–198 (2001). https://doi.org/10.1016/S0378-3820(01)00187-4
10. Manouchehri, H.R., Rao, K.H., Forssberg, K.S.E.: Triboelectric charge, electrophysical properties and electrical beneficiation potential of chemically treated feldspar, quartz and wollastonite. Magn. Electr. Sep. **11**, 9–32 (2002). https://doi.org/10.1155/2002/46414
11. Xing, Q., de Wit, M., Kyriakopoulou, K., Boom, R.M., Schutyser, M.A.I.: Protein enrichment of defatted soybean flour by fine milling and electrostatic separation. Innov. Food Sci. Emerg. Technol. **50**, 42–49 (2018). https://doi.org/10.1016/j.ifset.2018.08.014
12. Xing, Q., Utami, D.P., Demattey, M.B., Kyriakopoulou, K., de Wit, M., Boom, R.M., Schutyser, M.A.I.: A two-step air classification and electrostatic separation process for protein enrichment of starch-containing legumes. Innov. Food Sci. Emerg. Technol. **66**, 102480 (2020). https://doi.org/10.1016/j.ifset.2020.102480
13. Dance, A.D., Morrison, R.D.: Quantifying a black art: the electrostatic separation of mineral sands. Miner. Eng. **5**, 751–765 (1992). https://doi.org/10.1016/0892-6875(92)90244-4
14. Wills' Mineral Processing Technology. Elsevier (2016)
15. Manouchehri, H.R., Rao, K.H., Forssberg, K.S.E.: Review of electrical separation methods. Min. Metall. Explor. **17**(1), 23–36 (2000). https://doi.org/10.1007/BF03402825
16. Dwari, R.K., Rao, K.H.: Dry beneficiation of coal—a review. Miner. Process. Extr. Metall. Rev. **28**, 177–234 (2007). https://doi.org/10.1080/08827500601141271
17. Bada, S.O., Falcon, R.M.S., Falcon, L.M.: The potential of electrostatic separation in the upgrading of South African fine coal prior to utilization-a review. J. S. Afr. Inst. Min. Metall. **110**, 691–702 (2010)
18. Li, J., Xu, Z., Zhou, Y.: Application of corona discharge and electrostatic force to separate metals and nonmetals from crushed particles of waste printed circuit boards. J. Electrostat. **65**, 233–238 (2007). https://doi.org/10.1016/j.elstat.2006.08.004
19. Butcher, D.A., Rowson, N.A.: Electrostatic separation of pyrite from coal. Magn. Electr. Sep. 19–30 (1995)
20. Dascalescu, L., Iuga, A., Morar, R.: Corona–electrostatic separation: an efficient technique for the recovery of metals and plastics from industrial wastes. Magn. Electr. Sep. 241–255 (1993)
21. Lottering, J.M., Aldrich, C.: Online measurement of factors influencing the electrostatic separation of mineral sands. J. S. Afr. Inst. Min. Metall. 283–290 (2006)
22. Edward, D., Holtham, P.N., Kojovic, T.: The motion of mineral sand particles on the roll in high tension separators. Magn. Electr. Sep. 69–85 (1995)
23. Kelly, E.G., Spottiswood, D.J.: The theory of electrostatic separations: A review part III. The separation of particles. Miner. Eng. **2**, 337–349 (1989). https://doi.org/10.1016/0892-6875(89)90003-4
24. Svoboda, J.: Separation of particles in the Corona-discharge field. Magn. Electr. Sep. **4**, 173–192 (1993)
25. Ziemski, M., Holtham, P.N.: Particle bed charge decay behaviour under high tension roll separation. Miner. Eng. **18**, 1405–1411 (2005). https://doi.org/10.1016/j.mineng.2005.02.013
26. Germain, M., Lawson, T., Henderson, D.K., MacHunter, D.M.: The application of new design concepts in high tension electrostatic separation to the processing of mineral sands concentrates. Proc. Heavy Miner. 101–107 (2003)
27. Jafari, M., Chegini, G., Akmal, A.A.S., Rezaeealam, B.: A roll-type corona discharge–electrostatic separator for separating wheat grain and straw particles. J. Food Process Eng. **42**, e13281 (2019). https://doi.org/10.1111/JFPE.13281
28. Ore Kinetics. Available online: https://www.orekinetics.com.au

29. Kelly, E.G., Spottiswood, D.J.: The theory of electrostatic separations: a review Part I. Fundamentals. Miner. Eng. **2**, 33–46 (1989). https://doi.org/10.1016/0892-6875(89)90063-0
30. Kelly, E.G., Spottiwood, D.J.: The theory of electrostatic separations: a review part II. Particle charging. Miner. Eng. **2**, 193–205 (1989). https://doi.org/10.1016/0892-6875(89)90040-X
31. Dwari, R.K., Mohanta, S.K., Rout, B., Soni, R.K., Reddy, P.S.R., Mishra, B.K.: Studies on the effect of electrode plate position and feed temperature on the tribo-electrostatic separation of high ash Indian coking coal. Adv. Powder Technol. **26**, 31–41 (2015). https://doi.org/10.1016/j.apt.2014.08.001
32. Dwari, R.K., Hanumantha Rao, K.: Non-coking coal preparation by novel tribo-electrostatic method. Fuel **87**, 3562–3571 (2008). https://doi.org/10.1016/j.fuel.2008.05.029
33. Dwari, R.K., Hanumantha Rao, K.: Fine coal preparation using novel tribo-electrostatic separator. Miner. Eng. **22**, 119–127 (2009). https://doi.org/10.1016/j.mineng.2008.05.009
34. Ma, F., Tao, Y., Xian, Y., Zhang, M.: Effects of pulverized coal modification on rotary tribo-electric separation. Energy Sources, Part A Recov. Util. Environ. Eff. 1–13 (2020). https://doi.org/10.1080/15567036.2020.1772908
35. He, J., Yao, Y., Lu, W., Long, G., Bai, Q., Wang, H.: Cleaning and upgrading of coal-series kaolin fines via decarbonization using triboelectric separation. J. Clean. Prod. **228**, 956–964 (2019). https://doi.org/10.1016/j.jclepro.2019.04.329
36. Bittner, J.D., Hrach, F.J., Gasiorowski, S.A., Canellopoulus, L.A., Guicherd, H.: Triboelectric belt separator for beneficiation of fine minerals. Procedia Eng. **83**, 122–129 (2014). https://doi.org/10.1016/j.proeng.2014.09.021
37. Manouchehri, H.R., Rao, K.H., Forssberg, K.S.E.: Changing potential for the electrical beneficiation of minerals by chemical pretreatment. Min. Metall. Explor. **16**(3), 14–22 (1999). https://doi.org/10.1007/BF03402814
38. Wang, J., Dai, H.X., Luo, L.: The application status of Triboelectric separation and its progress. Adv. Mater. Res. **734–737**, 1114–1118 (2013). https://doi.org/10.4028/www.scientific.net/AMR.734-737.1114
39. Matsusaka, S., Maruyama, H., Matsuyama, T., Ghadiri, M.: Triboelectric charging of powders: a review. Chem. Eng. Sci. **65**, 5781–5807 (2010). https://doi.org/10.1016/j.ces.2010.07.005
40. Bada, S.O., Falcon, L.M., Falcon, R.M.S., Du Cann, V.M.: Qualitative analysis of fine coals obtained from triboelectrostatic separation. J. S. Afr. Inst. Min. Metall. **112**, 55–62 (2012)
41. Hansen, L., Wollmann, A., Weers, M., Benker, B., Weber, A.P.: Triboelectric charging and separation of fine powder mixtures. Chem. Eng. Technol. **43**, 933–941 (2020). https://doi.org/10.1002/ceat.201900558
42. Ma, F., Tao, Y., Liu, J., Xian, Y.: Flow field and particle motion characteristics of rotary triboelectric separator based on CFD simulation. Int. J. Coal Prep. Util. **00**, 1–22 (2020). https://doi.org/10.1080/19392699.2020.1847094
43. Landauer, J., Foerst, P.: Triboelectric separation of a starch-protein mixture—impact of electric field strength and flow rate. Adv. Powder Technol. **29**, 117–123 (2018). https://doi.org/10.1016/j.apt.2017.10.018
44. Li, T.X., Ban, H., Hower, J.C., Stencel, J.M., Saito, K.: Dry triboelectrostatic separation of mineral particles: a potential application in space exploration. J. Electrostat. **47**, 133–142 (1999). https://doi.org/10.1016/S0304-3886(99)00033-9
45. Arsentyev, V.A., Vaisberg, L.A., Ustinov, I.D., Gerasimov, A.M.: Perspectives of reduced water consumption in coal cleaning. In: XVIII International Coal Preparation Congress: 28 June–01 July 2016 Saint-Petersburg, Russia, pp. 1075–1081 (2016). https://doi.org/10.1007/978-3-319-40943-6_168
46. Eichas, K., Schonert, K.: A double drum separator for triboelectric separation of very fine materials.pdf. 18th International Mineral Processing Congress, pp. 417–423 (1993)
47. Hwang, Y., Gon, B., Bae, K., Seok, H.: Mechanism and performance of a dry particle separator using an elastic drum. Int. J. Miner. Process. **125**, 34–38 (2013). https://doi.org/10.1016/j.minpro.2013.09.006
48. Chen, J., Honaker, R.: Dry separation on coal—silica mixture using rotary triboelectrostatic separator. Fuel Process. Technol. **131**, 317–324 (2015). https://doi.org/10.1016/j.fuproc.2014.11.032

49. Youjun, T., Ling, Z., Dongping, T., Yushuai, X., Qixiao, S.: Effects of key factors of rotary triboelectrostatic separator on efficiency of fly ash decarbonization. Int. J. Min. Sci. Technol. (2017). https://doi.org/10.1016/j.ijmst.2017.06.004
50. Tao, D., Al-Hwaiti, M.: Beneficiation study of Eshidiya phosphorites using a rotary triboelectrostatic separator. Min. Sci. Technol. **20**, 357–364 (2010). https://doi.org/10.1016/S1674-5264(09)60208-8
51. Tao, D., Sayed-Ahmed, A., Li, Q., Honaker, R.: Dry fine coal cleaning using rotary triboelectrostatic separator (RTS). SME Annual Meeting and Exhibit (2010)
52. Chen, Q., Wei, L.: Coal dry beneficiation technology in china: the state-of-the-art. China Particuol. **1**, 52–56 (2003). https://doi.org/10.1016/s1672-2515(07)60108-0
53. Liu, Q., Bai, J., Gu, W., Peng, S., Wang, L., Wang, J., Li, H.: Leaching of copper from waste printed circuit boards using Phanerochaete chrysosporium fungi. Hydrometallurgy **196**, 105427 (2020). https://doi.org/10.1016/J.HYDROMET.2020.105427
54. Aman, F., Morar, R., Köhnlechner, R., Samuila, A., Dascalescu, L.: High-voltage electrode position: a key factor of elctrostatic separation efficiency. IEEE Trans. Ind. Appl. **40**, 905–910 (2004). https://doi.org/10.1109/TIA.2004.827813
55. Zhang, S., Forssberg, E.: Optimization of electrodynamic separation for metals recovery from electronic scrap. Resour. Conserv. Recycl. **22**, 143–162 (1998). https://doi.org/10.1016/S0921-3449(98)00004-4
56. Samuila, A., Urs, A., Iuga, A., Morar, R., Aman, F., Dascalescu, L.: Optimization of corona electrode position in roll-type electrostatic separators. IEEE Trans. Ind. Appl. **41**, 527–534 (2005). https://doi.org/10.1109/TIA.2005.844859
57. Zhang, G., He, Y., Wang, H., Zhang, T., Yang, X., Wang, S., Chen, W.: Application of triboelectric separation to improve the usability of nonmetallic fractions of waste printed circuit boards: removing inorganics. J. Clean. Prod. **142**, 1911–1917 (2017). https://doi.org/10.1016/j.jclepro.2016.11.093
58. Dodbiba, G., Shibayama, A., Sadaki, J., Fujita, T.: Combination of triboelectrostatic separation and air tabling for sorting plastics from a multi-component plastic mixture. Mater. Trans. **44**, 2427–2435 (2003). https://doi.org/10.2320/matertrans.44.2427
59. Jeon, H.S., Park, C.H., Kim, B.G., Park, J.K.: Development of triboelectrostaic separation technique for recycling of final waste plastic. Geosystem Eng. **9**, 21–24 (2006). https://doi.org/10.1080/12269328.2006.10541250
60. Urvantsev, A.I., Kashcheev, I.D.: Dry beneficiation of kyanite ores. Refract. Ind. Ceram. **54**, 166–168 (2013). https://doi.org/10.1007/s11148-013-9570-2
61. Tao, Y., Ding, Q., Deng, M., Tao, D., Wang, X., Zhang, J.: Electrical properties of fly ash and its decarbonization by electrostatic separation. Int. J. Min. Sci. Technol. **25**, 629–633 (2015). https://doi.org/10.1016/j.ijmst.2015.05.017
62. Wu, J., Li, J., Xu, Z.: Electrostatic separation for multi-size granule of crushed printed circuit board waste using two-roll separator. J. Hazard. Mater. **159**, 230–234 (2008). https://doi.org/10.1016/j.jhazmat.2008.02.013
63. Bedeković, G., Trbović, R.: Electrostatic separation of aluminium from residue of electric cables recycling process. Waste Manag. **108**, 21–27 (2020). https://doi.org/10.1016/j.wasman.2020.04.033
64. Silveira, A.V.M., Santana, M.P., Tanabe, E.H., Bertuol, D.A.: Recovery of valuable materials from spent lithium ion batteries using electrostatic separation. Int. J. Miner. Process. **169**, 91–98 (2017). https://doi.org/10.1016/j.minpro.2017.11.003
65. Yang, X., Wang, H., Peng, Z., Hao, J., Zhang, G., Xie, W., He, Y.: Triboelectric properties of ilmenite and quartz minerals and investigation of triboelectric separation of ilmenite ore. Int. J. Min. Sci. Technol. **28**, 223–230 (2018). https://doi.org/10.1016/j.ijmst.2018.01.003

Chapter 5
Sensor-Based Separation

5.1 Introduction

Modern technological developments are indispensable in mining and mineral processing to meet the global demands for metals and primary materials. However, along with rising ore deposit complexity, the ore grade is also decreasing. These cause an increase in the process energy cost, which is the most important challenge that the mining and mineral processing industry is currently facing. Undoubtedly employing innovative separation technologies would make processing procedures more efficient, sustainable, and cost-effective. In this context, automatic sensor-based sorting (ASBS) has been introduced as an appropriate and feasible option to overcome some of these technical challenges [1–3]. ASBS (also called ore sorting, sensor-based sorting, automated sorting, or electronic sorting) is a comparatively modern separation technique in which a sensor mechanically separates coarse particles (10–350 mm) based on particular physical and/or chemical characteristics [4]. This technology, as a modern alternative method for handpicking sorting (the oldest separation process technique [5]) was initially patented in 1928 [6, 7] and has been started industrially to be used in mining since the 1950s (Fig. 5.1) [8, 9].

A wide range of sensors has rapidly developed and advanced during recent years. Varied type of sensors has been utilized in the different steps of mining routes (Fig. 5.2). Recently, in the processing plants, what has been getting more attention is using sensors right before the grinding units for eliminating the waste portion of coarse particles within a feed. Rejection of coarse waste particles before the grinding could substantially decrease overall energy, reagents, and water consumption in downstream processes. ASBS systems have been developed and used to deal with this rejection. They could be well incorporated into mineral processing circuits to eliminate gangue coarse particle minerals before conventional processing treatment [10–12].

As tangible examples, right after the diamond ore has been concentrated by the heavy-dense medium separation (HDMS), ASBS has been utilized for all ultimate stages of its recovery [5]. Additionally, finding affordable alternative methods to

S. C. Chelgani and A. Asimi Neisiani, *Dry Mineral Processing*,
https://doi.org/10.1007/978-3-030-93750-8_5

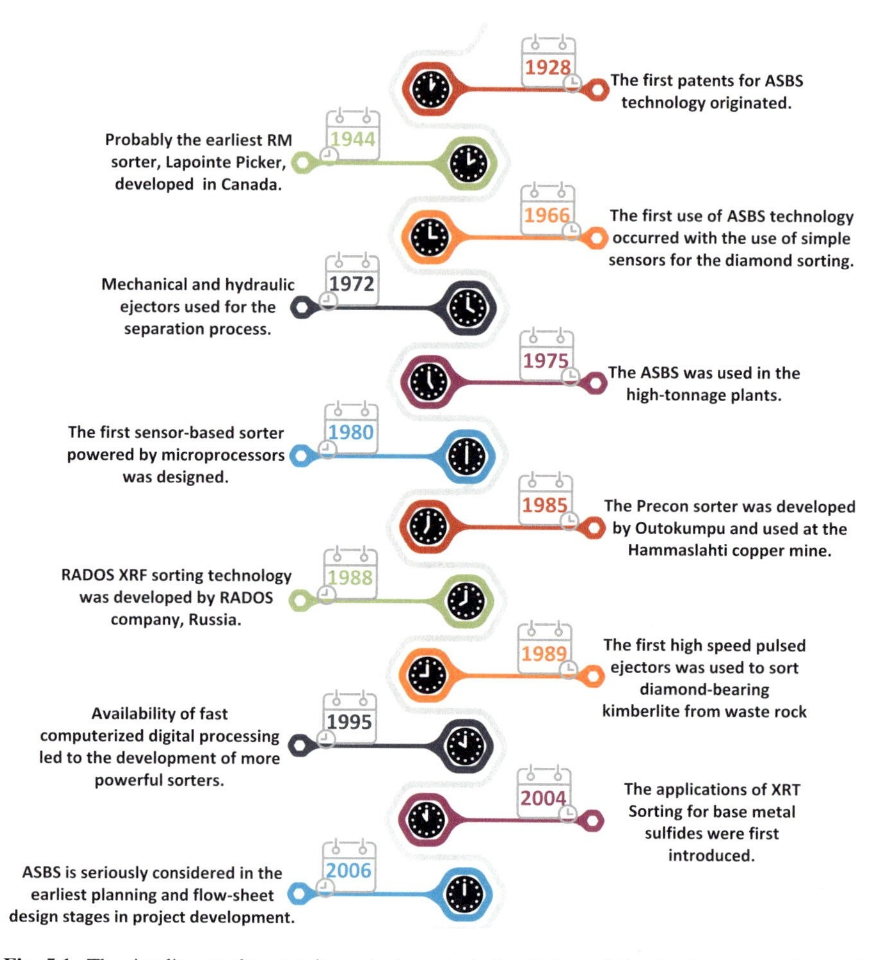

Fig. 5.1 The timeline on the most important events in the history of the development and use of this technology

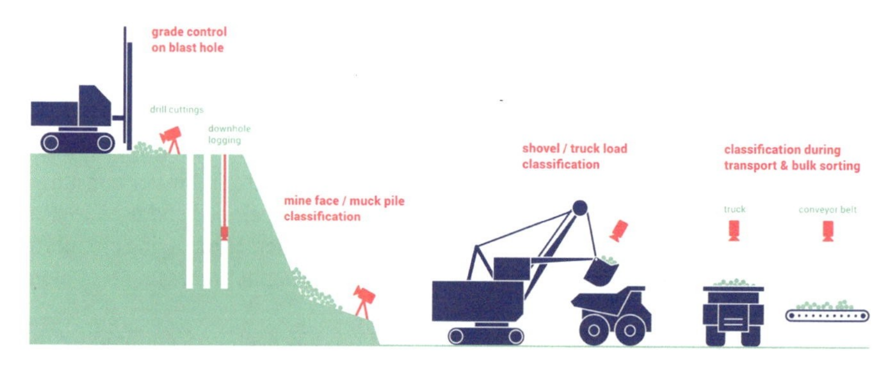

Fig. 5.2 Overview of opportunities for sensor-based material characterization in mining industry

separate the valuable grains effectively prior to the flotation process is under investigation in the zinc and lead industry. Some investigations have shown a great potential to substitute HDMS with ASBS techniques [2, 13].

5.2 Principles

Various sensors have been developed that can be used to identify different materials (Fig. 5.3) [14]. The electromagnetic spectrum is a phrase that refers to all known electromagnetic radiation wavelengths. Sensors are generated mainly based on the detection of these wavelengths. Currently, the detection of several features, including size, color, surface texture, shape, and density, is possible by sensors. According to the information obtained from manufacturers, approximately 50 various variables

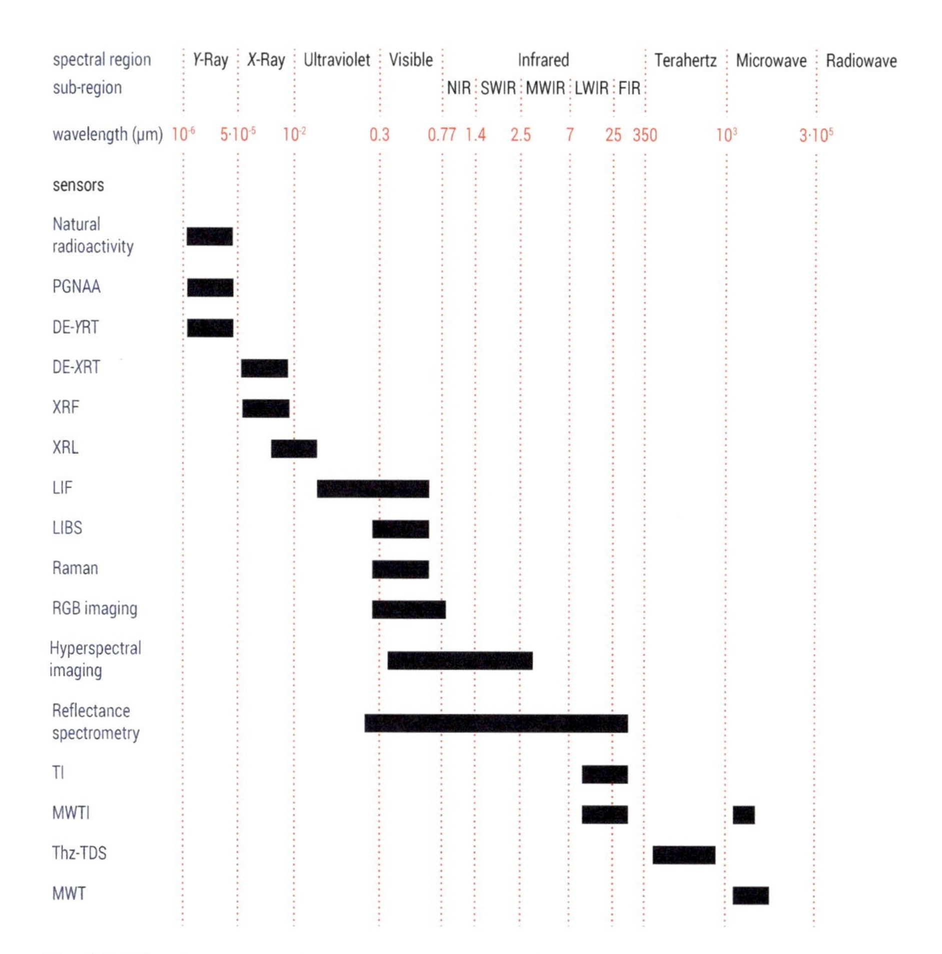

Fig. 5.3 The electromagnetic spectrum and wavelength ranges on which the different types of sensor technologies are based

could be utilized for mineral sorting [2]. After liberation, various sensors are able to recognize materials using their different properties. Examples of sensor-based separation in the mining and mineral processing industry comprise uranium ore sorting using natural radiometric (RM) detection [8], sorting diamonds based on X-Ray Luminescence (XRL) [15], and coal beneficiation based on with Dual-Energy X-Ray Transmission (DE-XRT) [13]. Moreover, sensors can be used as a raw material controller by installing over conveyor belts. Successful examples in which are RM sensors to calculate coal's ash content [16] and Prompt Gamma Neutron Activation Analysis (PGNAA) sensors to distinguish iron ore quality [17].

Two almost methodically identical sensor-based sorting procedures are called particle-by-particle sorting and bulk sorting (Fig. 5.4). In the former mode, particle-by-particle technique, materials are exposed to sensors in a single layer, and none of the particles touches each other. In the bulk sorting technique, particles enter the separator in layer-like groups because of the greater feeding rate. That is why valuable particles and low-grade surrounding particles are directed to the product part, maximizing the separation recovery. Consequently, the particle-by-particle sorting technique provides a higher-grade product in one run process. While the bulk sorting mode results in a higher recovery product. Despite the slight differences mentioned, they run in the same procedure (Fig. 5.5). Due to the technical similarity of the two

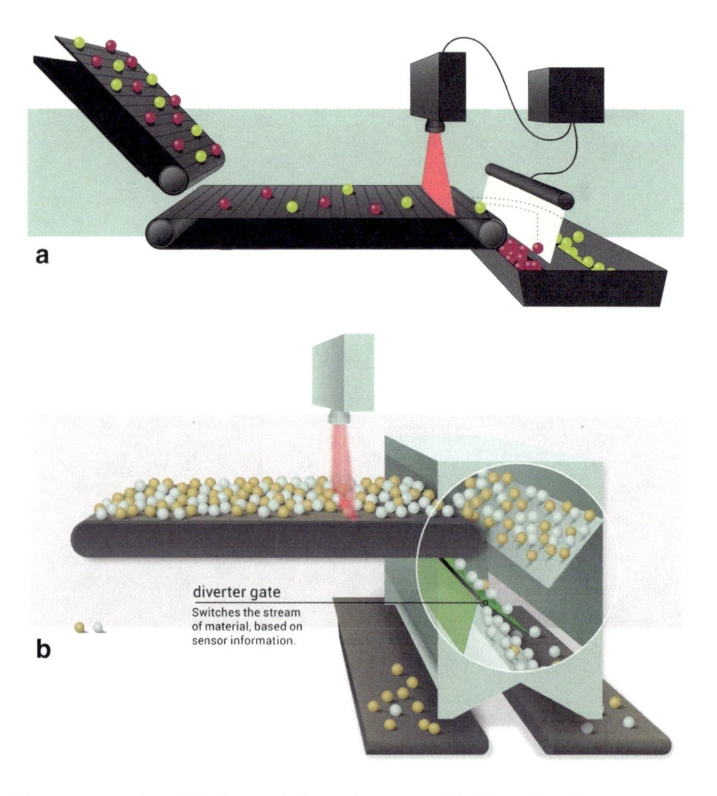

Fig. 5.4 The concept of particle-by-particle sorting (**a**) and bulk sorting (**b**)

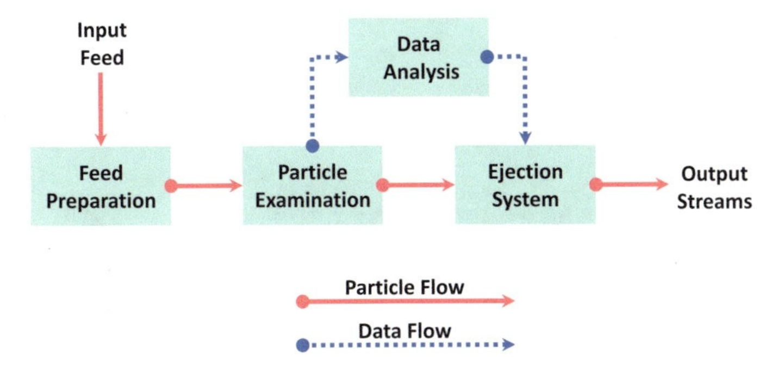

Fig. 5.5 The schematic operation of sensor-based sorting [20]

methods, both have been called particle-by-particle separation in some investigations [18, 19]. In principle, ASBS basically includes four interactive sub-steps (Fig. 5.5), namely:

1. Particle presentation;
2. Particle examination;
3. Data analysis;
4. Particle separation [20].

In the first step (Fig. 5.5), particle presentation is carried out after mono-layer feeding by either belt or chute-type system. Subsequently, sensors analyze the characteristics of particles in the particle examination step. After gathering the data, an analyzer categorized the particles (data analysis) based on their properties, and they will be sorted by an ejector (particle separation). In handpicking, the particle ejection or particle acceptation is done according to a yes/no decision. However, in the case of ASBS, the decision is made by assessing acquired information by the sensors using a sorting algorithm [21]. In the ejection step, several ejector types can be used, such as suction nozzles, mechanical ejectors, water jets, and pneumatic jet nozzles (the latter is the most effective and widely used) (Fig. 5.6) [5, 9].

It is clear that the efficiency of each major and minor stage directly impacts the efficiency of the whole ASBS process (Fig. 5.7). Material conditioning, applied before the sorting process begins, plays an important role in successful and effective separation. In the case of particle size, getting liberation and the lowest possible amount of fine particles, which is highly unfavorable to reach satisfactory efficiency, are imperative. The upper limit of particle size in the ASBS method is 350 mm lower limit of the particle size for most types of sensors is 0.5 mm. However, the liberation of low-grade ore is experienced quite below millimeters (mostly below 100 μm). However, from an economic point of view, the particle size of 10–20 mm is the optimum range for the ASBS process. With regard to this issue, it should be noted that the ratio of maximum size to minimum size should not be more than three [23, 24]. It was reported that using a single sensor for an extensive range of particle size would considerably decrease the sorting efficiency. From a technical point of view, the overall separation efficiency can be increased by simultaneously using two different types of sensors which have been arranged in series [2]. For

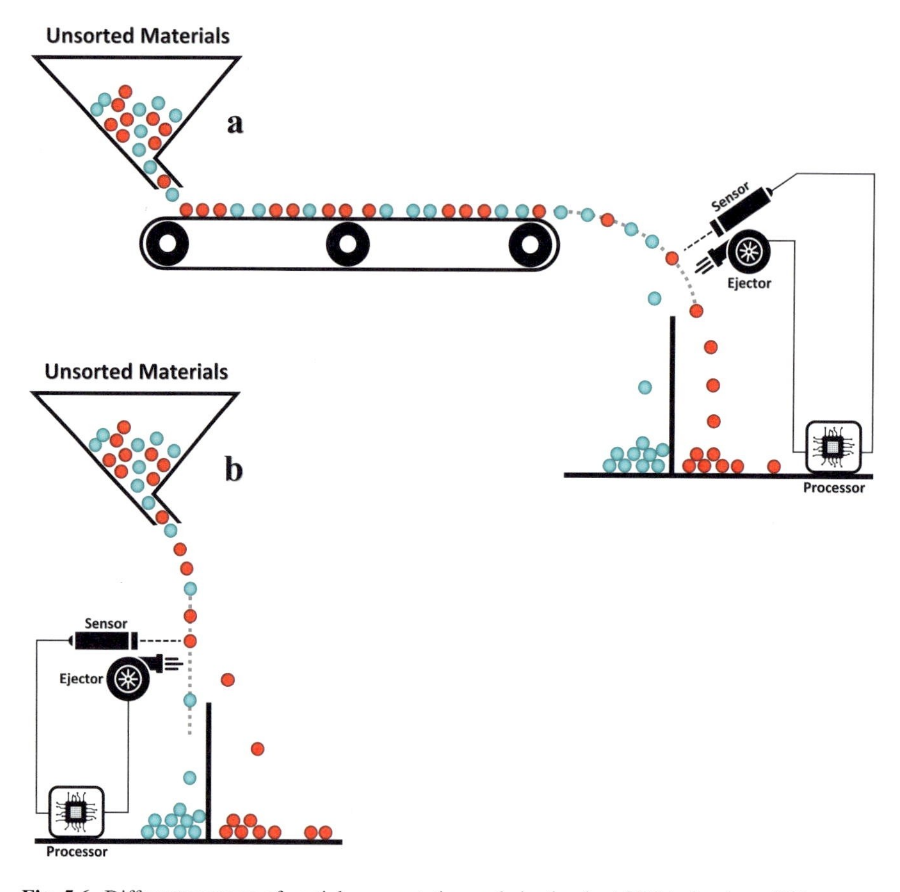

Fig. 5.6 Different systems of particle presentation and ejection in ASBS technology [22]

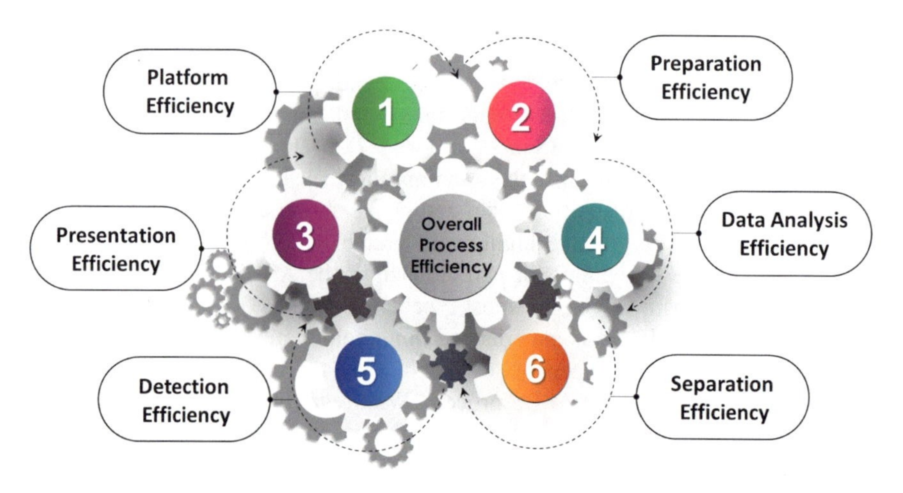

Fig. 5.7 The effect of ASBS sub-division efficiency on the overall effeciency

example, a double-sensor sorter, which consists of both optical and inductive, which can measure more than one feature, has been used in a previous study [25]. This application which is more suitable for mineral processing has been validated by successfully separating a nickel/copper sulfide ore and an iron ore.

5.3 ASBS in Mineral Processing

In mineral processing, the application of ASBS is rapidly increased. The ASBS capital and operating expenditures are approximately half of the cost of conventional pre-concentration coarse particle methods (gravity separation techniques). In many cases, ASBS has replaced the HDMS even in countries with low labor costs [9]. However, the ASBS process is entirely based on particle size. The costs of the process and many other parameters are strongly dependent on the average particle size fed to the separator machine. In fact, the ore liberation degree determines the lower particle size limit in the mineral processing industry. The costs of each ton of the product determine the upper limit of particle size. On the other hand, the rapid growth of electronics and the great variety of sensors has made it possible for the mineral processing industry to have many options to choose for ASBS depending on the type of ores and their characteristics. In general, X-ray-based methods play a significant role in the ASBS for the industry. X-ray luminescence (XRL) is the first and most extensively applied sensor-based sorting technology for minerals (diamond concentration). However, there are various commercially available sensors for separating different types of minerals (Table 5.1) based on their physical and chemical features [5, 9, 26].

Besides being affordable, ASBS processes are mainly dry processes and comparatively semi-mobile technology, making them more interesting. On the other hand, detected parameters by ASBS are very important for the processing sector, which could be used in other separation sections [3, 27, 28]. The great variety of available sensors and the programming in various algorithms make ASBS a very complex technology but flexible in terms of applications. There are several advantages in using ASBS and gangue particle elimination (pre-concentration) for different mineral processing units and mines since less mass need to be slurred and treated. They can be summarized as follows [5, 29–32]:

- Minimizing the power and water consumption in the grinding process.
- Decreasing reagent consumptions in froth flotation.
- Reducing the thickening unit capacity demand and flocculant consumption particularly in the tailings.
- Simplifying process control by reducing materials handling necessities.
- Increasing the production capacity.
- Final product preparation or production of different qualities in the cases such as diamonds, industrial minerals, and precious stones.
- Ability to use this method in underground mining.
- Extending mine life by old waste re-processing.

Table 5.1 Sensing technology available for ASBS [9]

Technology	Material property	Applications
Radiometric (RM)	Natural gamma radiation	Radioactive ores
Prompt gamma neutron activation analysis (PGNAA)	Absorption and emission of prompt gamma rays	Ferrous metals
X-ray transmission (XRT)	Atomic density	Base metal ores Precious metal ores Coal Tungsten Industrial minerals Diamonds Scrap metals Gold ores Fluorite
X-ray fluorescence (XRF)	Compositional analysis	Diamonds Base sulfide ores Precious metal ores Chromite Manganese ores
X-ray luminescence (XRL)	Visual fluorescence	Diamond
Color Camera Detection (CCD)	Reflection Absorption Transmission Brightness Transparency	Base metal ores Precious metal ores Industrial minerals Diamonds Gemstones
Visual Spectrometry (VIS)	Reflection or Absorption of visible radiation	Industrial mineral ores Gemstones
Photometric (PM)	Monochromatic reflection Absorption	Precious metal ores Industrial minerals Diamonds Gemstones
Near infrared spectroscopy (NIR)	Reflection Absorption	Base metal ores Industrial minerals Talc Limestone Lithium ores Kimberlite
Infrared camera (IR)	Heat conductivity Heat dissipation	Base metal ores Industrial minerals Precious metal ores Graphite Coal
Thermal infrared (TR)	Microwave excitation and thermal infrared detection	Base metal ores Precious metal ores
Electromagnetic (EM)	Conductivity Permeability	Base metal ores Scrap metals
Microwave infrared (MW-IR)	Microwave absorption Heat conductivity	Base metal ores Carbonaceous materials
Laser triangulation (LT)	Hull detection (shape and form)	Base metal ores Ferrous metals Precious metal ores

(continued)

Table 5.1 (continued)

Technology	Material property	Applications
Laser-induced fluorescence (LIF)	Absorption of laser light and spontaneous light emission	Industrial minerals
Laser-induced breakdown spectrometry (LIBS)	Evaporation of matter	Industrial minerals

- Elimination of disadvantageous part of plant feed which hampers downstream processes to improve overall recovery and the quality of final products.
- Making the feed uniform to subsequent processes for specialized treatment according to various characteristics such as grades, contaminant level, or mineralogical composition makes the process easier to handle.
- Reducing the amount of wet treated mineral tailing followed by environmental benefits.
- Reducing the fine particles and difficult-to-landfill materials.
- Mobile and flexible installation.
- Comparatively low effort legislation.
- Diminishing the cutoff grade by the reduction of the operating cost.

In addition to all of the above, arid territory, environmentally sensitive areas, and places lacking adequate energy supply are also dictating the potential of using the ASBS technology. Large well-known companies are currently developing and producing sensor-based separation equipment (Table 5.2). Today, a high-tech European company named TOMRA Sorting GmbH is the world's leading designer, developer, and manufacturer of sorting equipment for the mining industry.

5.3.1 X-Ray Transmission (XRT)

Rapidly developing methods of sensor-based beneficiation make it possible to achieve acceptable technological and economic results for different types of minerals. Among all electromagnetic wavelengths mentioned (Table 5.1), X-ray methods were officially the first technology used widely in the mineral processing industry. The x-ray transmission (XRT) scanning technology is extensively utilized at airports for baggage inspection. CommoDas has introduced the XRT principle and then has developed a sensor system that is suitably adapted for sorting techniques [33]. Subsequently, the mining and mineral processing industry has gradually become interested in this sorting technology. This system works like a line-scan camera and records the x-rays features penetrating the various particles and then converts them into digital image information. Based on the information obtained and the defined algorithms, the required action is ordered to the ejector for mineral sorting. Since pre-concentration using XRT sensors is one of the most common processes utilized in mineral processing, several cases of its successful use have been reported in the case of various ores and minerals, including scrap metals [34], gold [20, 35], tin [36], porphyry copper [37], and tungsten [38].

Table 5.2 Major companies are developing sensor-based sorters

Region	Company
Europe	SINTEF COMEX TOMRA REDWAVE Aweta B.V. MAF Industries QSort Sortex Barca Machine Vision Odenberg Engineering
North America	Delta Technology Durand-Wayland Ensco FMC Food Technology Focused Technologies Key Technology Produce Sorters International Sunkist TTI/Exeter Engineering Woodside Electronics
Asia	Satake RADOS
North America	Applied Sorting Technology Compaq

Despite the comparatively short history of using sensor-based separation in coal beneficiation, the separation devices that are developed based on XRT-technology have successfully adsorbed worldwide attractions. DE-XRT (dual-energy x-ray transmission) is a promising technical innovation that coal beneficiation plants have widely used for processing. This technology combines high-energy and low-energy levels of x-ray to detect particles. In the case of coal, which is mainly composed of carbon, because of its lower atomic density, the DE-XRT sensor can differentiate it from ash. One of the major challenges that some coal washery plants face is the presence of torbanite along with coal. This type of oil shale contains valuable smokeless fuel, which can be pyrolyzed and converted into liquid fuel. A final product containing a coal-torbanite mixture is neither appropriate for firing power stations nor suitable for exporting coal. In this case, torbanite causes problematic contamination at coal mines and beneficiation plants. Torbanite and coal have similar densities. Therefore, conventional density sorting methods such as jigging or HDMS show poor selectivity. In this case, DE-XRT as an ASBS technology has been demonstrated that could clearly differentiate coal from shale and torbanite (Fig. 5.8). DE-XRT separation also could effectually be used for the pyritic sulfur reduction in lignite (Brown Coal). If mercury is also present in these sources, its level has also decreased significantly with the separation of pyritic sulfides locked up in lignite since there is a high positive correlation between pyrite and mercury [33, 39–41].

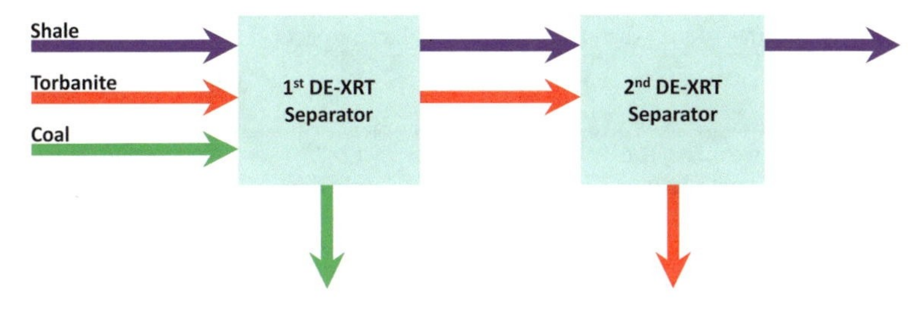

Fig. 5.8 Schematic diagram of the 2-stage DE-XRT separation of a coal-shale-torbanite mixture [41]

The promising results of using ASBS technology are not limited to coal beneficiation. HDMS could be used as a pre-concentration process for the lead and zinc oxidized ores. This conventional method is widely used around the world. However, it has several drawbacks and limitations (requires: high space -both height and footprint area, a great amount of water, a heating system in low-temperature areas, expensive heavy density materials, etc.), which result in a high operation cost, and low production capacity, particularly for the small units. However, using ASBS can reduce the investment cost up to 50% and the operating cost up to 20 times less. Table 5.3 shows the quantitative comparison made in the pre-processing process in both ASBS (Comex CXR) and HDMS methods [21].

One of the mega-projects that uses technology based on XRT-sensors is the greenfield plant at Umm Wu'al, Saudia Arabia, which is part of the Waad Al Shamal phosphate project, a joint venture between Saudi Arabian Mining Company Ma'aden. TOMRA developed and manufactured an XRT sorter for Ma'aden to be used in its new $560 million processing plant at Umm Wu'al. The plant can annually process around 13.5 million tons of raw material. TOMRA Sorting Mining's COM series machines sort inhomogeneous feed or critical moisture content. The XRT technology enables materials to be recognized and separated based on their average specific atomic density. This makes it possible to obtain a high purity level in sorted materials irrespectively of size, moisture, or contamination. These systems (Table 5.4) could sort more than 70% of the run-of-mine material by removing flint stones from the phosphate in order to reduce the silicon content. By this removal of waste

Table 5.3 Tests performed on dry pre-concentration of zinc and lead ore [21]

Option	Unit	HDMS	ASBS (Comex CXR)
Processing capacity	M ton/A	4.5–5	4.5–5
Water consumption	Lit/ton	1–2	0.01
Investment cost	Million USD	15	10–12
Operating cost	USD/ton	1–1.5	0.05–0.1

Table 5.4 The features of the Waad Al Shamal phosphate project [9]

Sensor type	Main challenge	Sorter capacity (ton per hour)	Size range (mm)	Main positive results
XRT	High wear because of flint present in phosphate	50	12–25	Reduction of wear in crushers and mills Reduction of the feed to the wet section of the plant Reduction of water and reagent consumption Reduction of fine tailings
		100	25–50	
		150	50–75	

material, the downstream process can be significantly downsized. The advantages gained will include a smaller plant footprint and considerable water, energy, and chemical reagents consumption per ton of final product [9, 42].

5.3.2 *X-Ray Fluorescence (XRF)*

ASBS methods that use XRF to separate particles have also recently been considered. The XRF technique considers the chemical composition of materials for sorting them and providing a pre-concentration. The XRF sorting method uses similar technology as the well-known XRF analyzer. Sensors utilize the interactions between surface particles and x-rays to determine the elemental composition of particles. In other words, this type of X-ray-based sensors sort minerals and rocks based on their chemical compositions by simultaneous measurement of the various element concentrations present on the particle's surface [43, 44]. The XRF technology can detect elements with an atomic number of more than 20, and then the system can sort particles accordingly. A pre-set pattern would be utilized to determine whether a certain particle is transported to the product or tailing section. During the main separation process, fines and misplaced particles are initially removed by a static grizzly in the ore processing plant. Subsequently, the remaining particles are transported to a vibrating feeder and several parallel channels or chutes. During the free-fall of the particles, the processes of detection and separation take place (Fig. 5.9). In the detection phase, X-rays strike the particle's surface, resulting in the emission of characteristic fluorescent X-rays from particles. The detector analyzes the characteristic X-rays emitted and their concentration, and the element composition present on the particle's surface would be calculated. After comparing the metal concentration of a specific particle with the pre-set sorting threshold value, the control unit would give orders to the mechanical ejector and eject the particle to either the product chute or the waste chute [45–47].

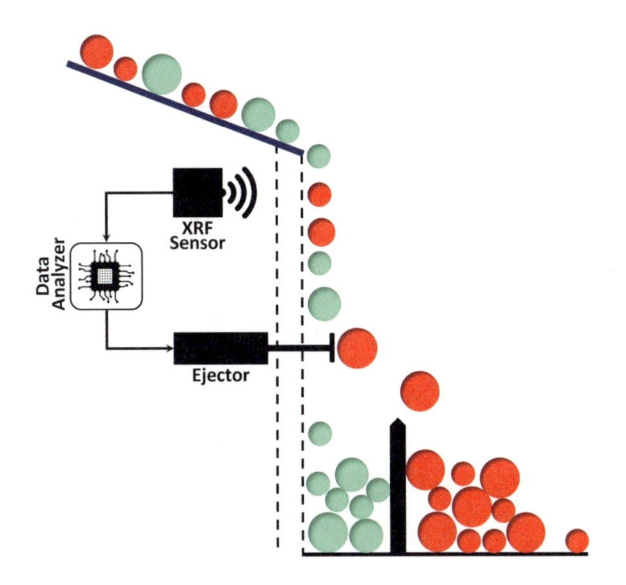

Fig. 5.9 Schematic diagram of XRF sorting process [44]

Although XRF reliability for mineral sorting is often questioned in various investigations due to the fact that X-Ray Fluorescence does not satisfactorily penetrate the particle, full-scale industrial XRF ore sorters are presently available from CommoDas and RADOS [48–50]. A full-scale XRF sorter by RADOS was installed in Svyatogor Copper-Zinc Mine, Russia (Fig. 5.10). XRF sorter manufactured by RADOS Company has an hourly capacity of 10–30 tons. Its capacity depends on ore density and particle size range. Each part is able to process and sort the particles with a size range of 20–250 mm, up to 8 particles per channel per second. This ore sorting technology consumes approximately 0.2–0.4 kWh energy per ton. Its detection limit is generally acceptable for most metals and even can increase for lighter elements. CommoDas Ultrasort has also manufactured XRF sorters with a capacity of up to 70 t/h. RADOS sorters have also been successfully used for the past three decades in Russia and Scandinavian countries. This dry separation technology is used for other minerals, such as precious metals, base metals, ferrous, and non-ferrous metals. For example, the results obtained from tests performed in captive manganese mines of Tata Steel, India, indicate a very high sorting accuracy [45, 46].

5.3.3 X-Ray Luminescence

One of the most extensively used ASBS for minerals is X-ray luminescence (XRL) technology, mainly used to concentrate diamonds. Using sensors in the mining and mineral processing industry began when the first patent was registered in 1928 to separate diamonds using XRL. This technology was practically used in the 1960s, where ASBS was considered for concentrating diamond and decreasing its losses

Fig. 5.10 Full-scale XRF sorter by RADOS installed in Svyatogor Copper-Zinc Mine, Russia

from the previously-applied grease-tables sorting. Another advantage of this technology compared to previous methods was the no need to use an attrition mill before the sorting step. This reduced energy consumption and the possibility of diamond breakage during the attrition process. XRL technology can be applied in almost all types of diamond except the high pure diamond (IIa type), which does not fluoresce in the exposure of X-ray radiation. The XRL in the diamond concentration procedure is mainly located right after the HDMS stage [8, 47]. Several investigations have been conducted for the XRL development. XRL sorting is limited to the class range of 1.25–32 mm because of self-absorption associated with larger diamonds [9, 51]. DeBeers and Alrosa developed sorting equipment called DebTech and Bourevestnik, respectively. Another provider was the Rio-Tinto Zinc technology development, which was followed by the UltraSort company. FlowSort is another supplier whose products are limited to small and medium scales. Because of the low capacity and the cascading flowsheets requirement, FlowSort approximately has been installed 1500 to 2000 XRL-based diamond sorters in total. Nowadays, the portable version of this modern-dry sorting technology is also available.

Table 5.5 The features of the renard diamond mine project [55]

Sensor type	Main challenge	Sorter capacity (ton per hour)	Size range (mm)
XRT	High	280	16–45
NIR	Non-kimberlite dilution	60	20–60
		150	60–180

5.3.4 Near Infrared (NIR)

Using the Near Infrared (NIR) spectroscopy technology is an innovative surface measurement technique that can be utilized for primary characterization, identification, and separation of favorable minerals from unwanted gangue during the ore sorting process. This technology, which is a single particle sorting method based on the absorbed/reflected radiation in the near-infrared (NIR) region, has been applied for decades in the laboratory scale to identify pure minerals. This method has its roots in several industries such as pharmaceuticals, recycling, and the food industry. The turning point of using this technology in the mining industry was conducted at RWTH Aachen University to evaluate the feasibility of using NIR to discriminate gangue phased from Skorpion zinc mine (an open-pit zinc-oxide mine located in Namibia which annually produces 150,000 tons of high-grade zinc). Skorpion ore comprises zinc-bearing minerals such as Hemimorphite, Hydrozincite, Sauconite, and Smithsonite. These types of minerals present diagnostic responses in the NIR region [52, 53]. Other studies have been conducted to investigate the feasibility of using this technology for the pre-concentration of other minerals such as copper-gold ore [54], talc [51], and kimberlite ore [52, 55].

One of the latest successful uses of the NIR-based sensors is in the Renard Diamond Mine, Canada. Before using NIR-based sensors, the XRT-based technology was used with large diamond recovery. After applying the secondary crusher, the fraction size of 16–45 mm was sorted by XRT sorter, and −16 mm particles fed to HDMS. The final product comprises the concentrate from the XRT sorter and the HDMS's concentrate. After April 2018, when NIR was utilized, there were four positive effects on the process:

- By ejecting hard rocks, diamond breakage was minimized.
- Energy consumption for crushing was decreased.
- The quality of the process's feed was upgraded.
- The Various circuit efficiencies were improved (Table 5.5).

5.4 Main Applications and Producers

In general, all these methods and systems have been used in different mineral processing investigations, which are summarized in Table 5.6. The representative producers of sensor-based separators are presented in Table 5.7.

Table 5.6 Summary of the sensor-based separators' application for various mineral beneficiation

Sensor	Feed	Location	Main challenge	Feed size	Feed rate	Main positive impacts	References
XRT	Tungsten	Mittersill Mine, Austria	Decreasing head-grade	16–30 mm 30–60 mm	25 tph 40 tph	– Decreased cut-o grade and increase in reserves – Increased flotation recovery – Reduction of fine tailings	[9]
XRT	Diamond	Karowe Mine, Botswana	High-density ore unsuitable for dense–medium separation (DMS) concentration	4–8 mm 8–14 mm 14–32 mm 30–60 mm 60–125 mm	30 tph 60 tph 120 tph 175t tph 250 tph	– Large diamond recovery – Replacement of DMS – Reduction of diamond breakage – Increased recovery – Reduction of specific operating costs	[9]
XRT	Phosphate	Umm Wu'al Mine, Saudi Arabia	High wear because of flint present in phosphate	12–25 mm 25–50 mm 50–75 mm	50 tph 100 tph 150 tph	– Reduction of wear in crushers and mills – Reduction of feed to the wet section of the plant – Reduction of water and reagent consumption – Reduction of fine tailings	[9]

(continued)

Table 5.6 (continued)

Sensor	Feed	Location	Main challenge	Feed size	Feed rate	Main positive impacts	References
XRT	Tin	San Rafael Mine, Peru	Declining head-grade, environmental liability for waste dump	6–14 mm 14–22 mm 22–32 mm 32–70 mm	20 tph 40 tph 60 tph 100 tph	– Reduction of cut-of-grade and increase in reserves – Increased flotation recovery – Reduction of fine tailings – Elimination of an environmental liability	[36]
XRT NIR	Diamond	Renard Mine, Canada	High non-kimberlite dilution	16–45 mm 20–60 mm 60–180 mm	280 tph 60 tph 150 tph	– Large diamond recovery – Increased recovery – Decreased specific operating costs – Increased plant capacity	[9]

(continued)

Table 5.6 (continued)

Sensor	Feed	Location	Main challenge	Feed size	Feed rate	Main positive impacts	References
XRT	Cu-Zn-Sn ore	Olkusz-Pomorzay Mine, Poland	The need to reduce water consumption as well as reduce high operating costs such as heavy medium consumption	20–60 mm	4500–5000 Ton $Year^{-1}$	– No need for water and heavy medium – More flexible operation – More environmentally friendly operation – In terms of the reduced investment and operating cost (by 15–20 times), it can dramatically improve the mine operation regarding economic aspects	[21]
DE-XRT	Bastnaesite	Artificial Samples	Mechanical processing of rare-earth ore types typically requires very high-energy consuming steps such as grinding to liberate valuable minerals. In addition, occasionally harmful reagents are used in processes such as flotation to separate the contained minerals	125–1000 μm	Laboratory scale	– Results show that samples with bastnaesite grades of 0.5 wt% were distinguishable from samples with no valuable content	[56]

(continued)

Table 5.6 (continued)

Sensor	Feed	Location	Main challenge	Feed size	Feed rate	Main positive impacts	References
Optical Sorter	Coal (lignite)	Different regions of Turkey	Feed does not respond to heavy medium separation because of the insufficient density difference between various particles	0–150 mm	Laboratory Scale	– Optical sorting can efficiently achieve a significant separation depending on the color differences between ash bearing content and coal – Reducing water consumption and dewatering cost – Reduce operating costs due to not requiring expensive heavy medium	[57]
Optical Sorter	Iron ore tailings	Turkey	Investigating the feasibility of hematite recovery from iron ore tailings	0–38 mm	Laboratory Scale	– The results showed that a salable hematite concentrate could be obtained with a high recovery from low intensity magnetic separator tailings by using two stage optical sorting Recovery of 96.43% Fe 51.78% Mass rejected 31.52%	[58]

(continued)

Table 5.6 (continued)

Sensor	Feed	Location	Main challenge	Feed size	Feed rate	Main positive impacts	References
XRT	silicate zinc ore	Brazil	Investigating the feasibility of pre-concentration of silicate zinc ore by ASBS	8–50 mm	Laboratory Scale	– It is attainable to recover around 93% of Zn in the concentrate and discard almost 30% of the mass as coarse waste	[59]

Table 5.7 The main sensor-based separator producers

Equipment	Manufacturer	Location
Sensor-based separators	Applied Sorting Technology	USA
	Aweta B.V.	Netherlands
	Barca Machine Vision	Multinational
	COMEX	Norway
	Compaq	USA
	Delta Technology	USA
	Durand-Wayland	USA
	Ensco	USA
	FMC Food Technology	USA
	Focused Technologies	USA
	Key Technology	USA
	MAF Industries	Multinational
	Odenberg Engineering	Ireland
	Produce Sorters International	USA
	QSort	Multinational
	RADOS	Russia
	REDWAVE	Austria
	Satake	Japan
	SINTEF	Norway
	Sortex	UK
	Sunkist	USA
	TOMRA	Norway
	TTI/Exeter Engineering	USA
	Woodside Electronics	USA

References

1. Knapp, H., Neubert, K., Schropp, C., Wotruba, H.: Viable applications of sensor-based sorting for the processing of mineral resources. ChemBioEng Rev. **1**, 86–95 (2014). https://doi.org/10.1002/cben.201400011
2. Mijał, W., Baic, I., Blaschke, W.: Modern methods of dry mineral separation—Polish experience. Lect. Notes Civ. Eng. **109**, 407–425 (2021). https://doi.org/10.1007/978-3-030-60839-2_21
3. Robben, C.: Characteristics of sensor-based sorting technology and implementation in mining, RWTH Aachen University, Department of Mineral Processing; Tomra Sorting Solutions/mining (2014)

4. Wotruba, H., Harbeck, H.: Sensor-based sorting. In: Ullmann's Encyclopedia of Industrial Chemistry. Wiley-VCH Verlag GmbH & Co. KGaA, Weinheim, Germany (2010)
5. Wills' Mineral Processing Technology. Elsevier (2016)
6. Salter, J.D., Wyatt, N.P.G.: Sorting in the minerals industry: past, present and future. Miner. Eng. **4**, 779–796 (1991). https://doi.org/10.1016/0892-6875(91)90065-4
7. Metallurgical separator (1926)
8. Sivamohan, R., Forssberg, E.: Electronic sorting and other preconcentration methods. Miner. Eng. **4**, 797–814 (1991). https://doi.org/10.1016/0892-6875(91)90066-5
9. Robben, C., Wotruba, H.: Sensor-based ore sorting technology in mining—past, present and future. Minerals **9** (2019)
10. Kolacz, J.: Advanced sorting technologies and its potential in mineral processing. AGH J. Min. Geoeng. **36**, 39–48 (2012)
11. Lessard, J., De Bakker, J., McHugh, L.: Development of ore sorting and its impact on mineral processing economics. Miner. Eng. **65**, 88–97 (2014)
12. Cutmore, N.G., Eberhardt, J.E.: The future of ore sorting in sustainable processing (2002)
13. Weatherwax, T.; Integrated mining and preconcentration systems for nickel sulfide ores. The University OF British Columbia (2007)
14. Workman, Jr., J., Weyer, L.: Practical Guide and Spectral Atlas for Interpretive Near-Infrared Spectroscopy. CRC Press (2012)
15. Society, C.M., Sul, L., Branch, S.: New methods and instruments in mining. **43**, 113–122 (2007)
16. Taylor, P.M., Cooke, A., Knight, C.B.: Natural Gamma for the On-line Measurement of the Ash Content of Conveyed Coal–25 Years of Success (2013)
17. Kurth, H.: Geoscan elemental analyzer for optimising plant feed quality and process performance. SAG Conferr. 1–11 (2015)
18. Wills, B.A., Finch, J.A.: Sensor-based Ore Sorting. Wills' Miner. Process. Technol. 409–416 (2016)
19. Napier-Munn, T.J., Morrell, S., Morrison, R.D., Kojovic, T.: Mineral comminution circuits: their operation and optimisation (1996)
20. von Ketelhodt, L.F.: Beneficiation of Witwatersrand type gold ores by means of optical sorting (2012)
21. Kolacz, J.: Advanced separation technologies for pre-concentration of metal ores and the additional process control. E3S Web Conf. **18** (2017)
22. De Jong, T.P.R.: Automatic sorting of minerals. In: Proceedings of the IFAC Proceedings Volumes (IFAC-PapersOnline); IFAC Secretariat, vol. 37, pp. 441–446 (2004)
23. Küppers, B., Parrodi, J.C.H., Lopez, C.G., Pomberger, R., Vollprecht, D.: Potential of sensor-based sorting in enhanced landfill mining. Detritus **8**, 24–30 (2019)
24. Christopher, R., Matthew, K., von, L.K.: Potential of sensor-based sorting for the gold mining industry, pp. 191–200 (2013)
25. Fitzpatrick, R.S.: The Development of a Methodology for Automated Sorting in the Minerals Industry (2008)
26. Robben, C., Condori, P., Takala, A.: Sensor-based ore sorting at San Rafael Mine. Int. Miner. Process. Conf. Moscow IMPC Counc. Russ. Acad. Sci. (2018)
27. Schindler, I.: Simulation-based comparison of Cut-and-Fill Mining with and without Preconcentration. Diplom Thesis, RWTH Aachen Univ. Aachen, Ger. (2003)
28. Dammers, M., Schropp, C., Martens, P.N., Rattmann, L.: Synergies and potentials of near-to-face processing—an integrated study on the effects on mining processes and primary resource efficiency, pp. 283–297 (2013)
29. Bamber, A., Klein, B., Morin, M., Scoble, M.: Reducing Selectivity in Narrow-Vein Mining through the Integration of Underground, pp. 1–12 (2004)
30. Wotruba, H.: Sensor sorting technology—is the minerals industry missing a chance? IMPC 2006—Proceedings of 23rd International Mineral Processes Congress, pp. 21–29 (2006)
31. Li, L., Li, G., Li, H., Li, G., Zhang, D., Klein, B.: Bench-scale insight into the amenability of case barren copper ores towards XRF-based bulk sorting. Miner. Eng. **121**, 129–136 (2018). https://doi.org/10.1016/j.mineng.2018.02.023

32. Buxton, M., Benndorf, J.: The use of sensor derived data in optimization along the Mine-Value-Chain. 15th International ISM Congress, Aachen, 16–20 Sept 2013, pp. 324–336
33. Kolacz, J.: New high definition X-ray sorting system based on X-MINE detection technology. IOP Conf. Ser. Mater. Sci. Eng. **641** (2019). https://doi.org/10.1088/1757-899X/641/1/012028
34. Mesina, M.B., de Jong, T.P.R., Dalmijn, W.L.: Automatic sorting of scrap metals with a combined electromagnetic and dual energy X-ray transmission sensor. Int. J. Miner. Process. **82**, 222–232 (2007). https://doi.org/10.1016/j.minpro.2006.10.006
35. Kleine, C., Riedel, F., Von Ketelhodt, L., Murray, R.: XRT Sorting of Massive Quartz Sulphide Type Gold Ore, pp. 1–10 (2016)
36. Robben, C., Condori, P., Pinto, A., Machaca, R., Takala, A.: X-ray-transmission based ore sorting at the San Rafael tin mine. Miner. Eng. **145**, 105870 (2020). https://doi.org/10.1016/j.mineng.2019.105870
37. Roukema, M.: DE-XRT assessment on porphyry copper ore from Chile Mick Roukema, Delft University of Technology (2013)
38. Robben, M., Knapp, H., Dehler, M., Wotruba, H.: X-ray transmission sorting of tungsten ore. Opt. Charact. Mater. 245–258 (2013)
39. De Jong, T.P.R., Dalmijn, W.L., Kattentidt, H.U.R.: Dual Energy X-Ray Transmission Imaging for Concentration and Control of Solids, pp. 1542–1551 (2003)
40. Wieniewski, A., Szczerba, E., Nad, A., Luczak, R., Kolacz, J., Szewczuk, A.: Evaluation of the application possibilities of modern separation techniques for pre-concentration of the Zn-Pb ore, pp. 27–30 (2015)
41. Von Ketelhodt, L., Bergmann, C.: Dual energy X-ray transmission sorting of coal. J. S. Afr. Inst. Min. Metall. 371–378 (2010)
42. TOMRA secures major sorting system contract from Ma'aden. Available online: https://newsroom.tomra.com/tomra-sorting-mining-secures-major-contract-for-maadens-saudi-arabia-phosphate-project/ (accessed on 25 Apr 2021)
43. Li, G., Klein, B., Sun, C., Kou, J.: Lab-scale error analysis on X-ray fluorecence sensing for bulk ore sorting. Miner. Eng. **164**, 106812 (2021). https://doi.org/10.1016/j.mineng.2021.106812
44. Fickling, R.S.: An introduction to the RADOS XRF ore sorter. In: Proceedings of the 6th Southern African Base Metals Conference; 6th Southern African Base Metals Conference (2011)
45. Mohanan, S., Saxena, G., Raghu Kumar, C., Naik, M., Kumar, A.: Use of Rados XRF sorters: Experience at TaTa Steel. In: Proceedings of INFACON XIII—13th International Ferroalloys Congress Efficient technologies Ferroalloy Industry, pp. 25–29 (2013)
46. Nadolski, S., Samuels, M., Klein, B., Hart, C.J.R.: Evaluation of bulk and particle sensor-based sorting systems for the New Afton block caving operation. Miner. Eng. **121**, 169–179 (2018). https://doi.org/10.1016/j.mineng.2018.02.004
47. Lessard, J., Sweetser, W., Bartram, K., Figueroa, J., McHugh, L.: Bridging the gap: Understanding the economic impact of ore sorting on a mineral processing circuit. Miner. Eng. **91**, 92–99 (2016). https://doi.org/10.1016/j.mineng.2015.08.019
48. Weltje, G.J., Tjallingii, R.: Calibration of XRF core scanners for quantitative geochemical logging of sediment cores: theory and application. Earth Planet. Sci. Lett. **274**, 423–438 (2008). https://doi.org/10.1016/j.epsl.2008.07.054
49. Hasan, A.R., Solo-Gabriele, H., Townsend, T.: Online sorting of recovered wood waste by automated XRF-technology: Part II. Sorting efficiencies. Waste Manag. **31**, 695–704 (2011). https://doi.org/10.1016/j.wasman.2010.10.024
50. Rasem Hasan, A., Schindler, J., Solo-Gabriele, H.M., Townsend, T.G.: Online sorting of recovered wood waste by automated XRF-technology. Part I: detection of preservative-treated wood waste. Waste Manag. **31**, 688–694 (2011). https://doi.org/10.1016/j.wasman.2010.11.010
51. Robben, A.M., Korsten, C., Pressler, N., Audy, P.L., Robben, M., Korsten, C., Audy, P.L., Pressler, N.: Theory and operational experience of NIR sorting in the Talc industry. Sens. Based Sorting **2012**, 41–42 (2012)
52. Mahlangu, T., Moemise, N., Ramakokovhu, M.M., Olubambi, P.A., Shongwe, M.B.: Separation of kimberlite from waste rocks using sensor-based sorting at Cullinan Diamond Mine. J. S. Afr. Inst. Min. Metall. **116**, 343–347 (2016)

53. Robben, M., Buxton, M., Dalmijn, W., Wotruba, H., Balthasar, D.: Near-infrared spectroscopy (NIRS) sorting in the upgrading and processing of skorpion non-sulfide zinc ore. XXV International Mineral Processing Congress 2010, IMPC 2010, vol. 2, pp. 1179–1186 (2010)
54. Phiri, T., Glass, H.J., Mwamba, P.: Development of a strategy and interpretation of the NIR spectra for application in automated sorting. Miner. Eng. **127**, 224–231 (2018). https://doi.org/10.1016/j.mineng.2018.08.011
55. Holl, I., Feldman, V., Zampini, J., Cunningham, R.: The comissioning and start-up of Quebec's first Diamond mine - Stornoway's Renard Mine (2019)
56. Neubert, K., Wotruba, H.: Investigations on the detectability of rare-earth minerals using dual-energy X-ray transmission sorting. J. Sustain. Metall. 3–12 (2017)
57. Gulcan, E., Gulsoy, O., Çelik, İ.B., Olgun, Z., Karaoğuz, S.: Investigation of dry coal beneficiation with optical sorter. In: XVIII International Coal Preparation Congress: 28 June–01 July 2016 Saint-Petersburg, Russia, pp. 1155–1160 (2016)
58. Aghlmandi, ahad Optical Sorting of Iron Ore Tailings. XV BMPC Sozopol, Bulg
59. José, D., Bergerman, M.G., Young, A.S., Petter, C.O.: Pre-concentration potential evaluation for a silicate zinc ore by density and sensor-based sorting methods. REM Int. Eng. J. **72**, 335–343 (2019). https://doi.org/10.1590/0370-44672018720155

Chapter 6
Mixed Methods

6.1 Introduction

The necessity and importance of reducing water consumption and the need to improve the performance of some dry separators have always forced the beneficiation methods to improve the efficiency of dry processing methods. Integrating different dry separation methods and developing a new separator could be considered as a possible alternative. Two of the most well-known of these separator devices are FGX compound dry separator and magnetically stabilized fluidized bed, which will be discussed in this chapter.

6.2 FGX Compound Dry Separator

Fuhe Ganfa Xuan mei (FGX) (developed in China) is a commercially available dry separator that employs an inclined vibrating plate and fluidization airflow to separate various particles [1, 2]. FGX principally is a density-based technology with a specific air table type and comprises a perforated plate, a vibrating system, three air chambers, and a suspension section (Fig. 6.1) [3]. The perforated plate, on which the separation process takes place, has been transversely and longitudinally suspended and possessed several riffles on its surface. A blower supplies air steam that causes the feed on the plate to be fluidized. Simultaneously, a vibrator system imparts a spiral rotational motion to materials, leading to sliding them in the longitudinal direction. In a comparatively wide particle size range (top-to-bottom ratio of approximately 10:1), a distinct stratification occurs based on the material density. The stratification on the perforated separating plate is due to various forces, namely the upward fluidizing force, the vibration force, and the downward gravitational force of the materials [4, 5].

The transverse inclination of the separating deck causes the light particles accumulated in the upper part of the fluidized bed to slide through the low-density particle

S. C. Chelgani and A. Asimi Neisiani, *Dry Mineral Processing*,
https://doi.org/10.1007/978-3-030-93750-8_6

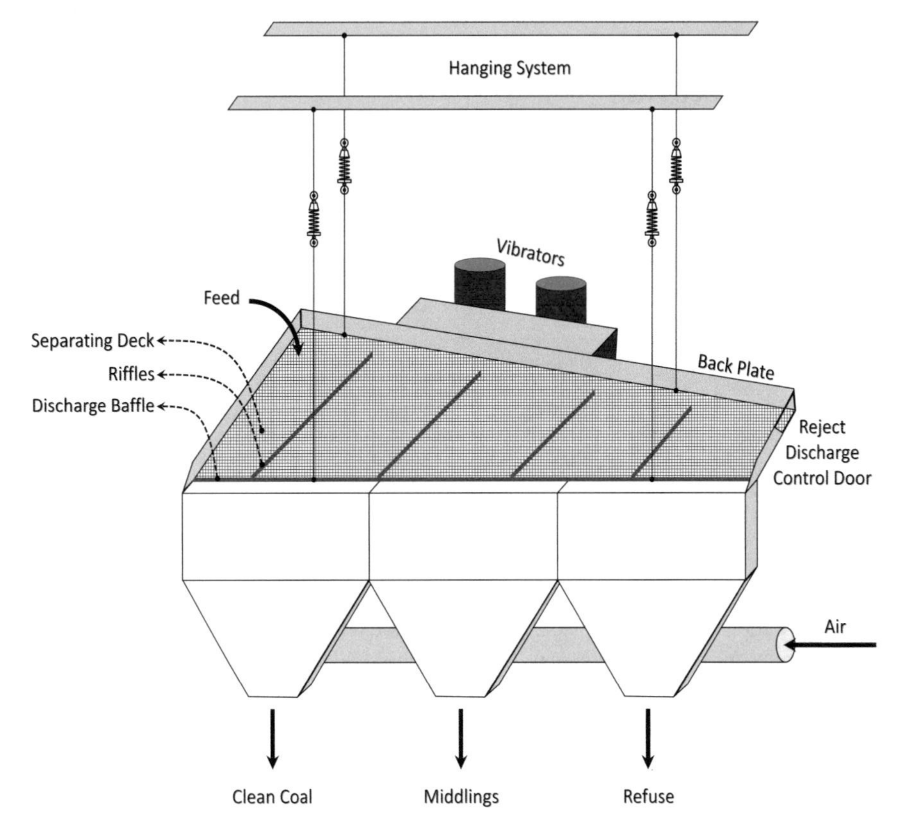

Fig. 6.1 The schematic structure of compound FGX dry separator [3]

partition continuously. On the other hand, denser particles concentrated in the bottom part of the stratified bed move toward the high-density outlet. Depending on the various particle characteristics, which fed to the separator, several products can be achieved. Typically, for dry coal processing, a FGX separator generates three output streams, including clean coal, middling, and refuse streams. All these output streams are collected along the front discharge section of the separator [6, 7].

In dry mineral processing, dust collecting is an important issue that has to be considered during separation. FGX separator is no exception from this necessity. For this purpose, a canopy in which a dust collector is embedded covers the perforated table. The negative pressure of the dust collector separates dust particles from the air blown into the system. The dust collector bags clean approximately 25% of air blown into the system during the separation process, whereas another 75% is circulated in a closed circuit [8]. Figure 6.2 illustrates a schematic diagram of the typical separation layout using a FGX separator [4].

FGX is a relatively new dry separator with remarkable efficiency and comparatively low costs. These properties have led to the extensive application of this method in China. Its first commercial application in the United States dates back to 2009 [6].

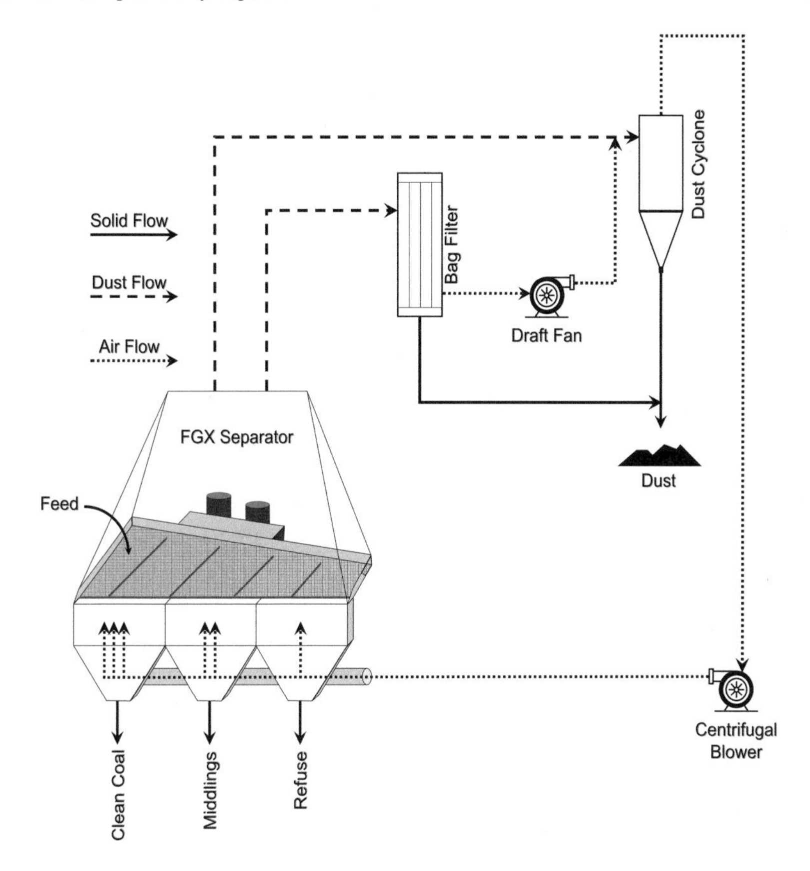

Fig. 6.2 A schematic diagram of dust handling in dry separation using a FGX separator [4]

This technology was developed by Tangshan Shenzhou Manufacturing Group and extensively utilized in coal beneficiation, more specifically to remove ash from coal (−80 mm size fraction) [9]. Several parameters can affect coal beneficiation process using FGX compound dry separator. The most important of these parameters are:

- Particle size range
- The moisture of particles
- The amount of 0–6 mm fraction in the input materials
- Vibration frequency of the separation deck
- Total ash content in the feed
- Total sulfur content in the feed
- The size of the biggest particles
- The number of middling particles in the feed
- Transverse and longitudinal angle of the separating deck
- Amount of air supplied to the separation zones [1, 10, 11].

Table 6.1 Approximate investment outlay and operating costs [10]

Type of process	Investment expenditures (dollar per ton h^{-1})	Operating costs (dollar per ton)
FGX separator	6200	0.50
Wet enrichment methods	13,000	1.95

Based on the effective factors, different operating variables have been examined to determine optimum conditions and enhance separation efficiency for various feeds [12–15]. The performance of FGX in separating various ranks of coals such as bituminous, sub-bituminous, and lignite is also investigated [16, 17]. The effect of different size fractions on the separation efficiency of high-sulfur content coal and the effect of various forces on particles at different heights of the bed have been explored [4, 18–20]. It is concluded that the particle size range should be from 6 to 75 mm to achieve an effective separation using a compound FGX separator. However, materials smaller than 6 mm may enhance the separation efficiency. It could be that − 6 mm particles would be required to have a fluidized bed with an autogenous dense medium. This would be led to improve the stratification process of the materials [17, 21, 22].

In addition to the advantages of this method, due to the lack of water usage, economic reasons have also led to increasing attention for using this machine. Economic efficiency analysis has indicated that investment expenditures and operating costs can be reduced if a coal beneficiation unit utilizes FGX instead of wet separation methods. As Table 6.1 shows, using FXG, compared to wet enrichment methods, reduces capital expenditures and operating costs by approximately 50% and 75%, respectively [10].

6.3 Magnetically Stabilized Fluidized Bed (MSFB)

In dry separation using air-dense medium fluidized beds, more specifically in coal beneficiation, the air bubbles disrupt the separation process of fine particles. This is why there is a limitation in the separation of fine size particles using this technology [23]. One way to solve this problem is to use the fluidized bed of magnetic materials which has been magnetically stabilized [24].

In an MSFB separator, when air velocity increases, large bubble formation is suppressed by the magnetic field, which is time-invariant and spatially uniform. In this process, fluidizing medium particles become axially oriented, and bed expansion occurs in a piston-like manner rather than a bubbling mechanism [25, 26]. Consequently, because of the magnetic field, the fluidizing bed is stabilized against the growth of disorders, such as bubble growth in the bed. MSFB mechanism reduces the solid or gas back mixing effect, which is usually a common phenomenon in

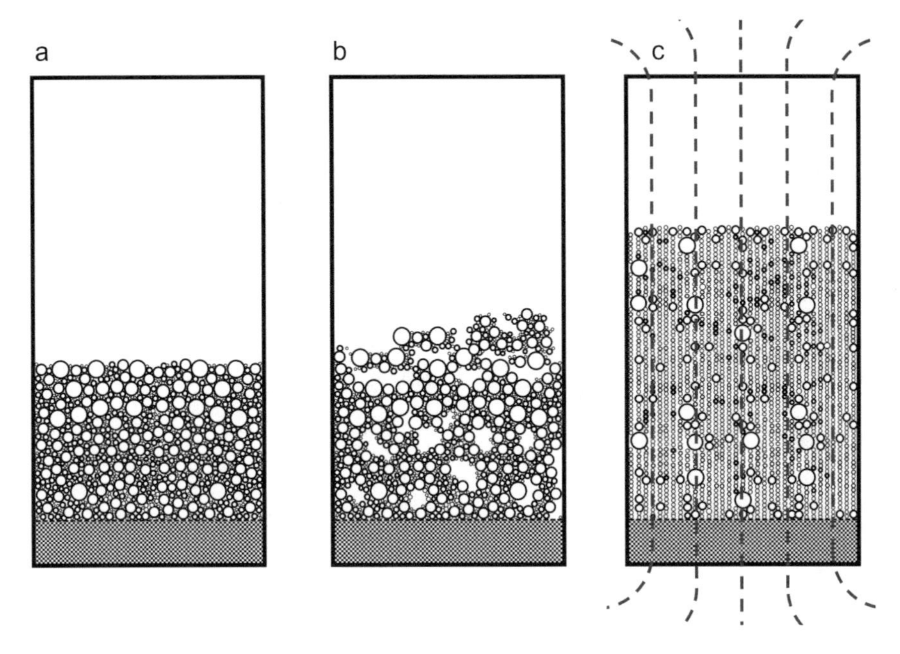

Fig. 6.3 Schematic of a fixed bed (**a**), bubbling bed (**b**), and MSFB (**c**) [24]

bubbling fluidized beds and facilitates the attainment of a more stable fluidized bed [9, 27, 28].

Figure 6.3 illustrates the fixed bed, typical bubbling bed, and magnetically stabilized fluidized bed. As it can be observed, there is no pulsation or bubbles when the fluidized bed becomes magnetically stabilized. This condition occurs when a magnetic field, which is collinear with fluidizing airflow, is applied to a bed of magnetic particles [24].

The magnetic stabilization generates a bubble-free bed in a fluid state, and therefore, the particles are easily transported. The operating velocity would be from the average minimum air velocity which is required for fluidizing the particles in the absence of the magnetic field, to the velocity that causes the bed to become a bubbling bed [29]. The magnetic particles arrange in chains corresponding to the imaginary lines of the magnetic field. Therefore, the maximum expansion rate in a magnetically stabilized fluidized bed is significantly greater than that of the common fluidized bed (Fig. 6.4). While particles are perfectly arranged with straight interstitial channels and create a more expanded fluidized bed in MSFB, the bed in a typical fluidized bed is randomly distributed [24].

The density of a magnetically stabilized fluidized bed consisting of magnetic medium and fine coal particles is:

$$\rho_b = \frac{r_v \rho_p}{(1 - r_m)(1 + i)}$$

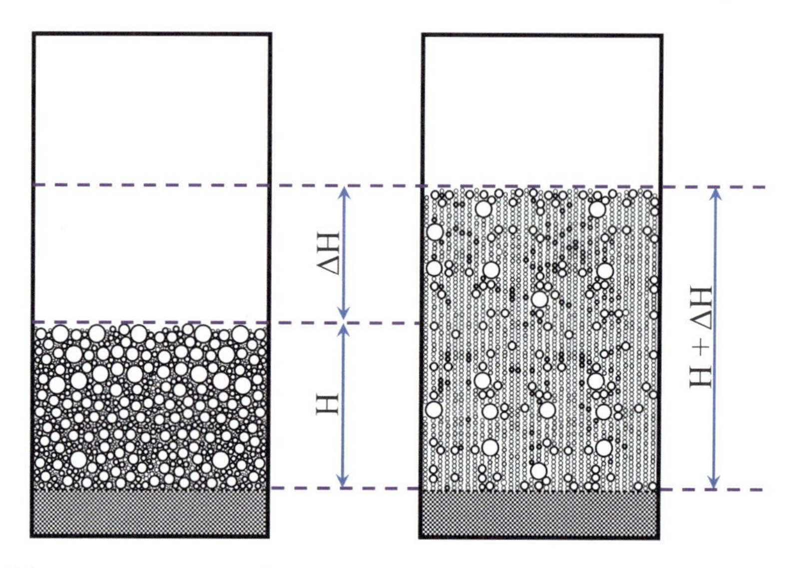

Fig. 6.4 The state of bed in an MSFB [24]

where

ρ_b density of the magnetically stabilized fluidized bed (kg m^{-3})
r_v volume fraction occupied by the magnetic medium
ρ_p bulk density of the magnetic medium (kg m^{-3})
r_m mass fraction of coal fines in the magnetically stabilized fluidized bed
i expansion ratio of the magnetically stabilized fluidized bed ($\Delta H/H$) [24]

The density of the magnetic medium in MSFB is from 1300 to 2200 kg m^{-3}, and ρ_b can be adjusted to the desired bulk density. After fine coal particles are fed into MSFB, they are separated based on density difference according to Archimedes' principle. The particles which possess a density greater than ρ_b sink into the bed, whereas those whose density is lower than ρ_b float on the top surface of MSFB [24].

References

1. Mijał, W., Niedoba, T., Polek, D.: Mathematical model of dry coal deshaling by using FGX vibrating air table. IOP Conf. Ser. Mater. Sci. Eng. **641** (2019). https://doi.org/10.1088/1757-899X/641/1/012025
2. Akbari, H., Zhang, B., Yang, F., Mohanty, M.K., Sayeh, M., Rahimi, S.: Application of neural network for modeling the coal cleaning performance of the FGX dry separator. Sep. Technol. Miner. Coal Earth Resour. 189–197 (2012)
3. Gombkötő, I., Madarász, T., Szűcs, P., Lakatos, J.: Dry cleaning, an affordable separation process for Deshaling Indian high ash thermal coal. In: XVIII International Coal Preparation Congress 28 June–01 July 2016 Saint-Petersburg, Russia (2016). https://doi.org/10.1007/978-3-319-40943-6

4. Zhang, B., Akbari, H., Yang, F., Mohanty, M.K., Hirschi, J.: Performance optimization of the FGX dry separator for cleaning high-sulfur coal. Int. J. Coal Prep. Util. **31**, 161–186 (2011). https://doi.org/10.1080/19392699.2011.574943
5. Ling, X., He, Y., Li, G., Tang, X., Xie, W.: Separation performance of different particle sizes using an industrial FGX dry separator. Int. J. Coal Prep. Util. **38**, 30–39 (2018). https://doi.org/10.1080/19392699.2016.1185418
6. Mijał, W., Baic, I., Blaschke, W.: Modern methods of dry mineral separation—Polish experience. Lect. Notes Civ. Eng. **109**, 407–425 (2021). https://doi.org/10.1007/978-3-030-60839-2_21
7. Dong, L., Wang, Z., Zhou, E., Wang, X., Li, G., Fan, X., Zhang, B., Duan, C., Chen, Z., Luo, Z., et al.: A novel dry beneficiation process for coal. Int. J. Coal Prep. Util. 1–21 (2019). https://doi.org/10.1080/19392699.2019.1692339
8. Baic, I., Blaschke, W., Góralczyk, S., Sobko, W., Szafarczyk, J.: Stanowisko badawcze do odkamieniania urobku węglowego metodą suchej separacji. Czas. Tech. Krak. Tow. Tech. 15–19 (2014)
9. Chen, Q., Wei, L.: Coal dry beneficiation technology in China: the state-of-the-art. China Particuol. **1**, 52–56 (2003). https://doi.org/10.1016/s1672-2515(07)60108-0
10. Blaschke, W., Baic, I.: FGX air-vibrating separators for cleaning steam coal—functional and economical parameters. IOP Conf. Ser. Mater. Sci. Eng. 641 (2019). https://doi.org/10.1088/1757-899X/641/1/012030
11. Blaschke, W.S., Szafarczyk, J., Baic, I., Sobko, W.: A study of the deshaling of Polish hard coal using an FGX unit type of air concentrating table. XVIII International Coal Preparation Congress, pp. 1143–1148 (2016)
12. Li, H., Luo, Z., Zhao, Y., Wu, W., Zhang, C., Dai, N.: Cleaning of South African coal using a compound dry cleaning apparatus. Min. Sci. Technol. **21**, 117–121 (2011). https://doi.org/10.1016/J.MSTC.2010.12.019
13. Kademli, M., Gulsoy, O.Y.: Investigation of using table type air separators for coal cleaning. Int. J. Coal Prep. Util. **33**, 1–11 (2013). https://doi.org/10.1080/19392699.2012.717566
14. Patil, D.P., Parekh, B.K.: Beneficiation of fine coal using the air table. Int. J. Coal Prep. Util. **31**, 203–222 (2011). https://doi.org/10.1080/19392699.2011.574948
15. Ghosh, T., Honaker, R.Q., Patil, D., Parekh, B.K.: Upgrading low-rank coal using a dry, density-based separator technology. Int. J. Coal Prep. Util. **34**, 198–209 (2014). https://doi.org/10.1080/19392699.2014.869934
16. Hacifazlioglu, H.: Production of merchantable coal from low rank lignite coal by using FGX and subsequent IR drying. Int. J. Coal Prep. Util. **40**, 418–425 (2020). https://doi.org/10.1080/19392699.2018.1450248
17. Honaker, R.Q., Saracoglu, M., Thompson, E., Bratton, R., Luttrell, G.H., Richardson, V.: Upgrading coal using a pneumatic density-based separator. Int. J. Coal Prep. Util. **28**, 51–67 (2008). https://doi.org/10.1080/19392690801934054
18. Kademli, M., Gulsoy, O.Y.: Influence of particle size and feed rate on coal cleaning in a dry separator. Physicochem. Probl. Miner. Process. **52**, 204–213 (2016). https://doi.org/10.5277/ppmp160118
19. Yu, X., Luo, Z., Gan, D.: Desulfurization of high sulfur fine coal using a novel combined beneficiation process. Fuel **254**, 115603 (2019). https://doi.org/10.1016/j.fuel.2019.06.011
20. Li-juan, S., Jian-zhong, C.: Analysis of material motion in compound dry separator. J. China Univ. Min. Technol. (2005)
21. Shen, L.: The function auto-medium serves in compound dry coal separator. Coal Prep. Technol. **2**, 45–48 (1998)
22. Li, G.M., Yang, Y.S.: Application of compound dry cleaning technique in China. Coal Process. Compr. Util. (2006)
23. Luo, Z., Zhao, Y., Chen, Q., Tao, X., Fan, M.: Separation lower limit in a magnetically gas–solid two-phase fluidized bed. Fuel Process. Technol. **85**, 173–178 (2004). https://doi.org/10.1016/S0378-3820(03)00175-9

24. Fan, M., Chen, Q., Zhao, Y., Luo, Z.: Fine coal (6–1 mm) separation in magnetically stabilized fluidized beds. Int. J. Miner. Process. **63**, 225–232 (2001). https://doi.org/10.1016/S0301-7516(01)00054-0
25. Hristov, J.Y.: Fluidization of ferromagnetic particles in a magnetic field Part 1: the effect of field line orientation on bed stability. Powder Technol. **87**, 59–66 (1996). https://doi.org/10.1016/0032-5910(95)03070-0
26. Mohanta, S., Rao, C.S., Daram, A.B., Chakraborty, S., Meikap, B.C.: Air dense medium fluidized bed for dry beneficiation of coal: technological challenges for future. Part. Sci. Technol. **31**, 16–27 (2013). https://doi.org/10.1080/02726351.2011.629285
27. Chen, Q., Wei, L.: Development of coal dry beneficiation with air-dense medium fluidized bed in china. China Particuol. **3**, 42 (2005). https://doi.org/10.1016/s1672-2515(07)60161-4
28. Yang, X., Zhao, Y., Luo, Z., Song, S., Chen, Z.: Fine coal dry beneficiation using autogenous medium in a vibrated fluidized bed. Int. J. Miner. Process. **125**, 86–91 (2013). https://doi.org/10.1016/j.minpro.2013.10.003
29. Hristov, J.Y.: Fluidization of ferromagnetic particles in a magnetic field Part 2: field effects on preliminarily gas fluidized bed. Powder Technol. **97**, 35–44 (1998). https://doi.org/10.1016/S0032-5910(97)03392-5

Printed in the United States
by Baker & Taylor Publisher Services